商业建筑美学

——解析商业街区与商业建筑设计 上册

Commercial Buildings Aesthetics

Design Analysis of Commercial Districts & Commercial Buildings (Vol.1)

商业建筑美学编委会 编著

商业广场&商业街&社区商业

Commercial Plazas & Commercial Streets & Community Commerce

华中科技大学出版社
http://www.hustp.com
中国·武汉

图书在版编目(CIP)数据

商业建筑美学：解析商业街区与商业建筑设计：全2册/商业建筑美学编委会编著. −武汉 ： 华中科技大学出版社，2015.6
ISBN 978−7−5609−8780−4

Ⅰ．①商… Ⅱ．①商… Ⅲ．①商业街−建筑设计②商业−服务建筑−建筑设计 Ⅳ．①TU984.13②TU247

中国版本图书馆CIP数据核字(2015)第017977号

商业建筑美学——解析商业街区与商业建筑设计：全两册

商业建筑美学编委会 编著

出版发行：华中科技大学出版社 (中国·武汉)

地　　址：武汉市武昌珞喻路1037号 (邮编：430074)

出 版 人：阮海洪

责任编辑：刘锐桢　　　　　　　　　　　　　　　　　责任监印：秦　英

责任校对：杨　睿　　　　　　　　　　　　　　　　　装帧设计：林国代

印　　刷：利丰雅高印刷 (深圳) 有限公司

开　　本：965 mm×1270 mm　1/16

印　　张：44

字　　数：352千字

版　　次：2015年6月第1版第1次印刷

定　　价：698.00元　 (全两册)

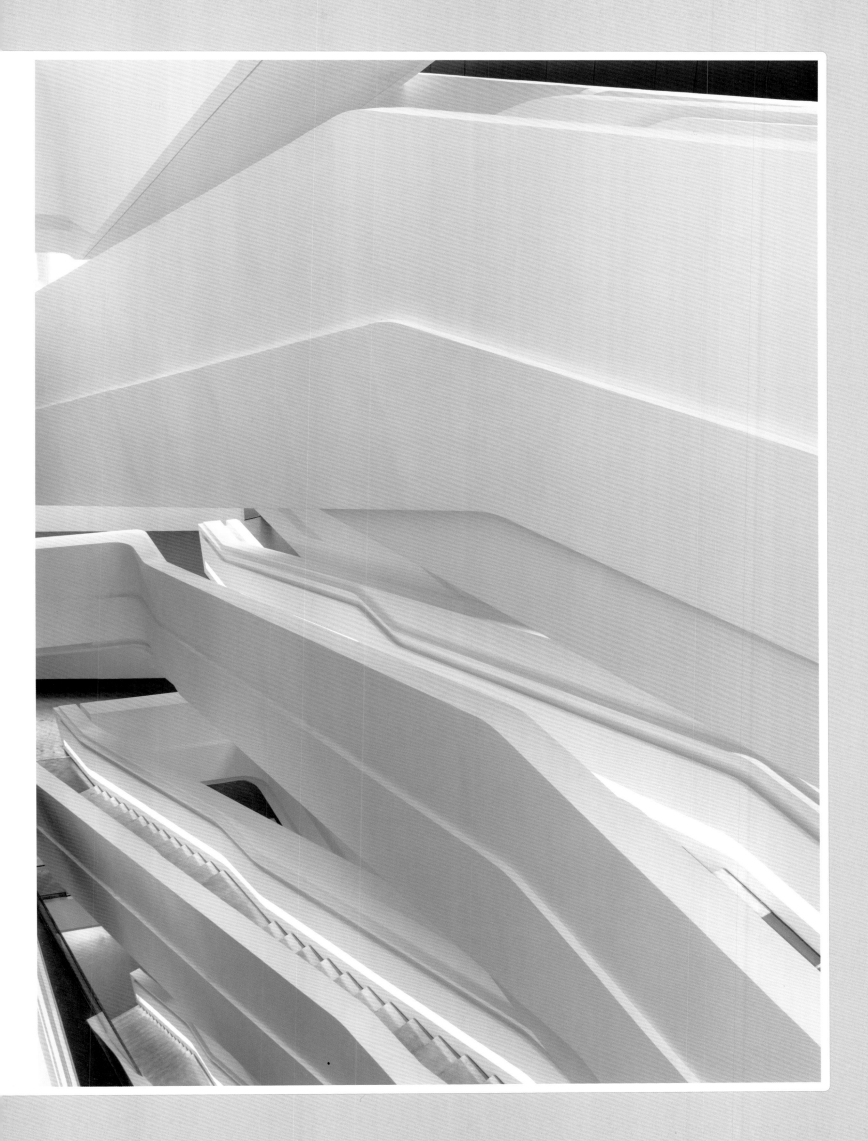

53.80 F.F.L.
OP OF PARAPET

47.50 F.F.L.
32 MEP ROOF

41.20 F.F.L.
31 MEP ROOF

37.00 F.F.L.
30 HIGH ZONE

32.80 F.F.L.
29

28.60 F.F.L.
28

24.40 F.F.L.
27

20.20 F.F.L.
26

16.00 F.F.L.
25

11.80 F.F.L.
24

07.60 F.F.L.
23

03.40 F.F.L.
22

9.20 F.F.L.
21

5.00 F.F.L.
20

0.80 F.F.L.
19 HIGH ZONE

6.60 F.F.L.
18 LOW ZONE

2.40 F.F.L.
17

8.20 F.F.L.
16

4.00 F.F.L.
15

9.80 F.F.L.
14

5.60 F.F.L.
13

1.40 F.F.L.
12

7.20 F.F.L.
11

3.00 F.F.L.
10

.80 F.F.L.
9

.60 F.F.L.
8

.40 F.F.L.
7

.20 F.F.L.
6

.00 F.F.L.
5 LOW ZONE

.60 F.F.L.
4

.20 F.F.L.
3

.80 F.F.L.
2

.40 F.F.L.
LOBBY / F&B

6.3 / 4.2 / 4.2 / 4.2 / 6.3 / 4.2 / 5.4 / 5.4 / 5.4 / 5.4

商业建筑是一个城市的名片，

是一个城市的形象。

美的建筑环境，

是一切商业价值的基础。

创造有独特功效的美，

并让它稀缺。

这，就是大商业时代的竞争要求。

Commercial buildings are a card of a city,

and the image of the city.

Beautiful architectural environment,

is the basis of all commercial value.

Create beauty of unique efficacy,

and make it scarce.

This is a requirement for competing in this grand business era.

在国内外很多奖项中,大家可能看不到太多的商业建筑。

这实际上是一个传统,并不是最近的现象。在建筑业实际存在着对商业建筑相当大的抵触情绪或抵触态度。

对于正统的建筑师来说,他们对商业建筑有抵触情绪的,他们认为真正的建筑是不能和商业化混为一谈的,真正的纯正的文化建筑是会被商业元素污染的。

从比较积极的角度来说,商业建筑最终的目的是追逐商业利益,而建筑师却有一个社会责任,就是维护公众的利益。从工作方式上来看也是不一样的,如果说做文化建筑,大家更多讨论的是一些想法、思路、概念、理念,那么在商业建筑设计里,第一步绝对不是这样开始,更多的是从功能空间的关系上入手,从这一点来说,它非常接近于城市设计,是非常理性的一个过程。在这个过程中很多情况下不是建筑师主导的,建筑师处于一个相对被动的位置,这也是很多建筑师不太愿意做商业建筑的一个重要原因。

本书从市场的角度,深入地阐述了每个项目从无到有的过程。每个项目中设计师都是遵循因地制宜、因势利导的设计原则,这种态度和立场是今天商业建筑领域里可喜的现象,是我们必须坚持的。希望能对大家有些启发!

In many awards at home and abroad, commercial buildings are not commonly seen.

This is actually a tradition, not a recent phenomenon. In fact, there exists considerable emotional resistance or attitudinal resistance in architectural field.

Orthodox master architects' emotional resistance toward commercial architecture is not only about emotion, but also about cultural background. They deem that true architecture can't be confused with commercialization, and true and pure cultural architecture is to be polluted by commercial elements.

From a positive perspective, the ultimate goal of commercial architecture is to after commercial interests. Also architects have a social responsibility of protecting public interests. When it comes to doing a cultural building, people talk more about ideas, concepts and theories. However, it's quite a different case in designing a commercial building where the first step is definitely not that in a cultural building, instead, the relation between function spaces is more likely to be taken as the start. From this perspective, commercial architecture design is quite similar to urban design, featuring a rational process. During this process, many conditions are not led by architects. On the contrary, architects are stuck in a relatively passive position. This is an important reason why many architects don't like doing commercial buildings.

Starting from the market perspective, the book does a deep analysis on each project developing from nothing. In each project, designers adhere to the principles of acting according to local conditions and making the best use of local conditions. This attitude and stance is a gratifying phenomenon in present commercial architecture field, and is what we should pass on. Hope the book will inspire you somehow.

目录

商业广场

Commercial Plazas

[012-031] 上海K11购物艺术中心
此项目充分体现了 K11品牌的核心价值：艺术、人文、自然，完美地为优质都会生活做出新提案。

K11 Art Mall, Shanghai
The project fully embodies the core value of K11 brand "art, culture, nature", and sets a new proposal for superior urban life.

[032-051] 上海环茂广场
上海环茂广场是一个由新型高端商业裙楼组成的综合项目。以两栋A级国际写字塔楼、豪华住宅塔楼及与地铁无缝连接为特征，上海环茂广场代表了这座城市综合商业项目的新未来。

iAPM
iAPM is a new high-end retail podium and the key feature for the Shanghai ICC mixed-use scheme. Featuring two Grade A international office towers, luxury residential tower and seamless connections to the MTR, Shanghai ICC with iAPM represents the new future for mixed-use retail schemes in the city.

[052-069] 哈尔滨哈西万达广场
哈尔滨哈西万达广场项目是万达集团斥巨资打造建设的哈尔滨第三座万达广场，是目前东三省规模最大的城市综合体。

Haxi Wanda Plaza, Harbin
The project is the third Wanda Plaza in Harbin built by Wanda Group, and it is the largest urban complex across the three northeast provinces of China currently.

[070-091] 抚顺万达广场
抚顺原为大清龙脉所在，采光顶以"龙行天下"为主题，室内设计元素以龙麟为根基，整条室内步行街宛如龙戏二珠。

Wanda Plaza, Fushun
The skylight takes dragon world as the mother theme and the interior design uses dragon scale as the basic element.

[092-119] 合肥砂之船艺术商业广场
本项目的规划理念为"一轴、一心、双环、六节点"。

Sasseur Art Commercial Plaza
The project features a planning concept of "one axis, one heart, two rings, six nodes".

[120-145] 圣塔莫尼卡广场
通过将传统商场的城市建设原则与零售设计有机结合，捷得事务所仔细而精心地将这个项目变成今日的城市建筑。

Santa Monica Place
Blending timeless urban principles that predate conventional malls with its organic approach to retail design, Jerde carefully and intricately wove the project into the existing city fabric.

[146-167] 阿布扎比阿尔达尔中央市场
该项目的设计灵感源自传统建筑，设计旨在彻底改造这个市场空间，为阿布扎比创造一个新的市民中心。

Aldar Central Market, Abu Dhabi
Inspired by the traditional architecture of the Gulf this scheme aims to reinvent the market place, giving the city a new civic heart.

[168-187] Maras Park
设计的一个关键点在于波浪状的天窗，通过打开的天窗可以捕捉到北边历史城市中心及南边Dibec山的景色。

Maras Park
A key element of the design, the undulating skylight opens to the sky above and captures multiple views to the North historic city center and to the Southern Dibec Mountains.

[188-209] 大邱彩色广场购物中心
受到体育馆标志性的曲线屋顶象征全球团结的启发，项目外墙覆盖了独特绚丽的马赛克图案，体现了世界多元文化。

Color Square Stadium Mall Daegu
Inspired by the stadium's iconic curved roof which symbolizes the bringing together of people from around the globe, the project's exterior walls are covered with unique and colorful mosaic patterns that are a metaphor for the world's numerous cultures.

CONTENTS

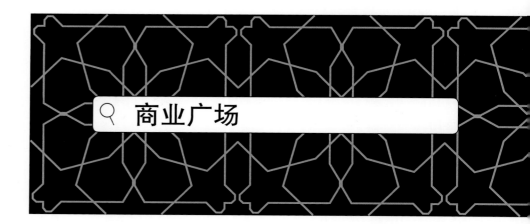

商业广场

Commercial Plazas

上海K11购物艺术中心
K11 Art Mall, Shanghai

地点：上海市　建筑面积：9 100平方米　业主：新世界　设计公司：柯凯建筑设计顾问（上海）有限公司　摄影：Charlie Xia
Location: Shanghai, China　Building Area: 9,100m²　Design Company: Kokaistudios　Photography: Charlie Xia

上海K11购物艺术中心
K11 Art Mall, Shanghai

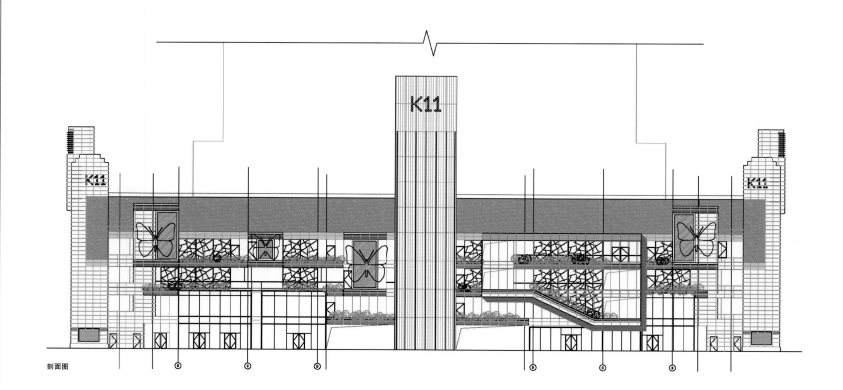

上海K11购物艺术中心
K11 Art Mall, Shanghai

剖面图

地点：上海市　建筑面积：9 100平方米　业主：新世界　设计公司：柯凯建筑设计顾问（上海）有限公司　摄影：Charlie Xia
Location: Shanghai, China　Building Area: 9,100m²　Design Company: Kokaistudios　Photography: Charlie Xia

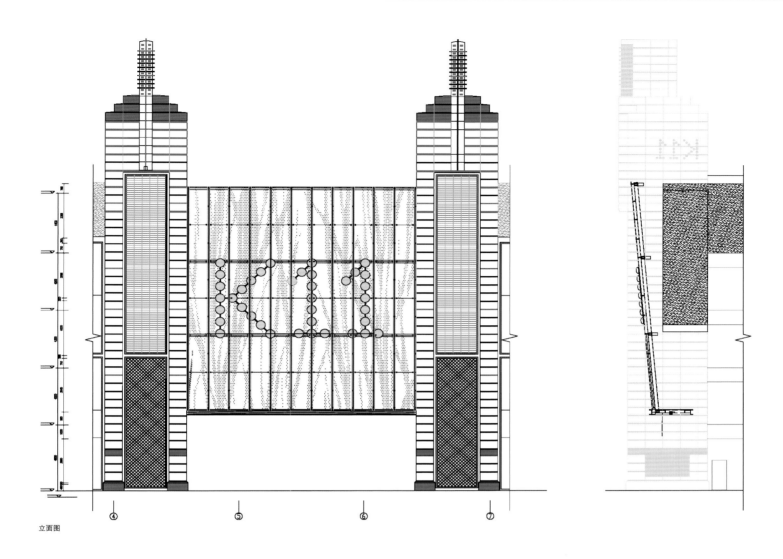

立面图

上海K11购物艺术中心
K11 Art Mall, Shanghai

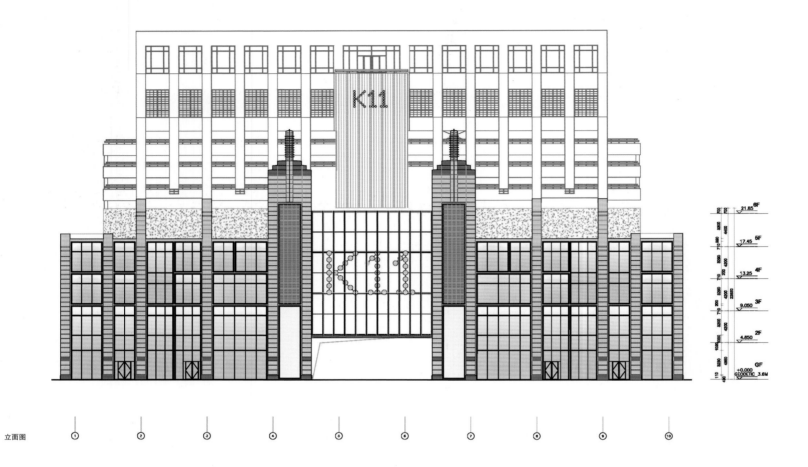

立面图

此项目坐落于上海市中心主要购物街淮海路的黄金地段，设计师以崭新的手法重新打造新世界大厦裙房。此栋20世纪80年代建成的地标性大楼为现今上海的商业建筑修缮工程项目树立了典范。Kokaistudios作为总建筑及室内设计的建筑事务所，负责统筹及监督35 500平方米的商业空间，加强了建筑物与其周围环境的关系。此项目充分体现了K11品牌的核心价值：艺术、人文、自然，完美地为优质都会生活做出新提案。

修缮裙房建筑立面将新旧结合，守旧与创新的平衡运用创造出绝佳的效果。在重视淮海路历史建筑和新世界塔楼原始设计的同时，满足在高密度环境下商场及租户对视觉通透性的需求。

商场的六个楼层在视觉上通过位于中庭的顶棚形成良好的联结及延续，这个由玻璃建造的有机形态的顶棚总面积达到280平方米。商场

出入口及循环的动线围绕中庭设计，将"想象之旅"和艺术展示、公共区域、高科技纵横错落地交织在一起，并借由生活元素与自然素材增添其人文内涵。

中庭拥有九层楼高的瀑布，透过自动电感系统随着气候条件调节水量，这是亚洲最高的户外水幕瀑布。超过2 000平方米的垂直绿化墙可将收集的雨水用于冷却建筑。

位于三、四层楼的餐饮区，拥有一座城市农场，可直接通往绿化完善的屋顶花园及停车场，是繁忙拥挤的商业街上难得的都市绿洲。面向中庭的走道玻璃窗在天气好时可向阳台滑动推开，从而将室内艺术空间转换成俯瞰中央庭院的通透场所。

透过位于地下三层的私人美术馆，K11将永久收藏的艺术品展示在商场内，无论是开幕、活动、讲座、设计竞赛或是展览，都让艺术走进生活，促进群众的参与。白天，阳光透过玻璃

顶棚照射到商场地下两层楼，而到了夜晚，地下商场内的灯光提供了一个由内向外的光源，将地面楼层照射得金碧辉煌。

此项目获得"LEED金级认证"。其在设计中加入了不同的节能措施以提高能源的使用效率，降低了"热岛效应"，减少了用水。可持续性材料的应用和无障碍公共交通的布局也是此次设计的重点。

地点：上海市　建筑面积：9 100平方米　业主：新世界　设计公司：柯凯建筑设计顾问 (上海) 有限公司　摄影：Charlie Xia

Location: Shanghai, China　Building Area: 9,100m²　Design Company: Kokaistudios　Photography: Charlie Xia

上海K11购物艺术中心
K11 Art Mall, Shanghai

地点: 上海市　建筑面积: 9 100平方米　业主: 新世界　设计公司: 柯凯建筑设计顾问 (上海) 有限公司　摄影: Charlie Xia
Location: Shanghai, China　Building Area: 9,100m²　Design Company: Kokaistudios　Photography: Charlie Xia

上海K11购物艺术中心
K11 Art Mall, Shanghai

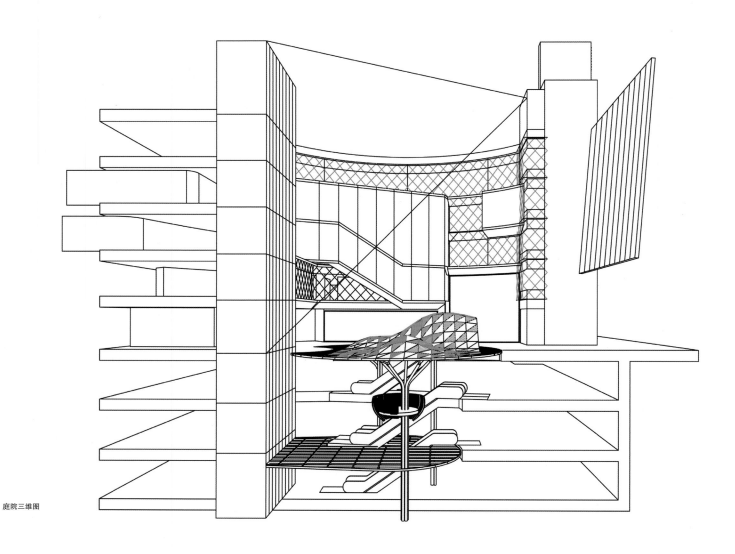

庭院三维图

地点：上海市　建筑面积：9 100平方米　业主：新世界　设计公司：柯凯建筑设计顾问 (上海) 有限公司　摄影：Charlie Xia
Location: Shanghai, China　Building Area: 9,100m²　Design Company: Kokaistudios　Photography: Charlie Xia

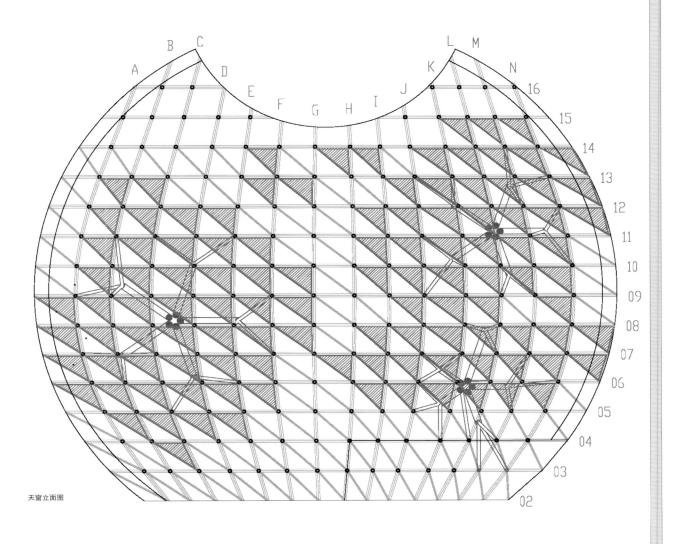

天窗立面图

Located at the gold section of Huaihai Road, a main shopping street in Shanghai downtown, the project reinvents the podium of New World Building through a brand-new approach, setting a great example for commercial building renovation projects in Shanghai. Kokaistudios is commissioned to coordinate and supervise the 35,500m² commercial space, and to strengthen the relation between the building and its surrounding environment. The project fully embodies the core value of K11 brand "art, culture, nature", and sets a new proposal for superior urban life.

The renovation design of the podium facade perfectly combines the old and the new, and finds a balance between conservativeness and innovativeness. While laying emphasis to the historical buildings on Huaihai Road and the original design of New World Building, the design also meets the mall and tenants' requirement of visual transparency in such a high-density environment.

The six floors of the mall are visually connected by the ceiling unfolded from floor in the atrium. The glass ceiling's area reaches to 280m². The mall exits and circulatory paths are arranged along the atrium, interweaving "imagination journey", art display, public area and high technology, and filling in cultural connotation by life elements and natural materials.

The nine-storey high waterfall in the atrium can be adjusted in terms of water flow. It is the highest outdoor water-screen waterfall in Asia. The vertical green wall that is higher than 2,000m can collect rain water to be used in cooling system.

The F&B facilities on the third and fourth floors have an urban farm which directly leads to the roof garden and parking area, and is a rare urban oasis in the busy and crowded commercial street. The glass window of the walkway that opens to the atrium can slide to balcony on a nice day, thus easily transforms the interior art space into a transparent space that can overlook the central courtyard.

With the private art gallery on the third underground floor, K11 displays its art collection and allows art to walk into people's life either through opening ceremony, event, lecture, design competition or exhibition. During the day, sunshine floods onto the second underground floor of the mall through the glass ceiling. At night, the lamp light in the underground mall offers an inside-out light source that lightens up the ground floors.

Except for the LEED gold certification, the project also draws in other energy-saving strategies, in the aim to improve energy efficiency, reduce heat island effect and decrease water use. Besides, sustainable materials and non-barrier public traffic are also the keys in the project.

上海K11购物艺术中心
K11 Art Mall, Shanghai

地点：上海市　建筑面积：9 100平方米　业主：新世界　设计公司：柯凯建筑设计顾问 (上海) 有限公司　摄影：Charlie Xia
Location: Shanghai, China　Building Area: 9,100m²　Design Company: Kokaistudios　Photography: Charlie Xia

地点：上海市　建筑面积：9 100平方米　业主：新世界　设计公司：柯凯建筑设计顾问 (上海) 有限公司　摄影：Charlie Xia
Location: Shanghai, China　Building Area: 9,100m²　Design Company: Kokaistudios　Photography: Charlie Xia

上海K11购物艺术中心
K11 Art Mall, Shanghai

地点：上海市　建筑面积：9 100平方米　业主：新世界　设计公司：柯凯建筑设计顾问 (上海) 有限公司　摄影：Charlie Xia
Location: Shanghai, China　Building Area: 9,100m²　Design Company: Kokaistudios　Photography: Charlie Xia

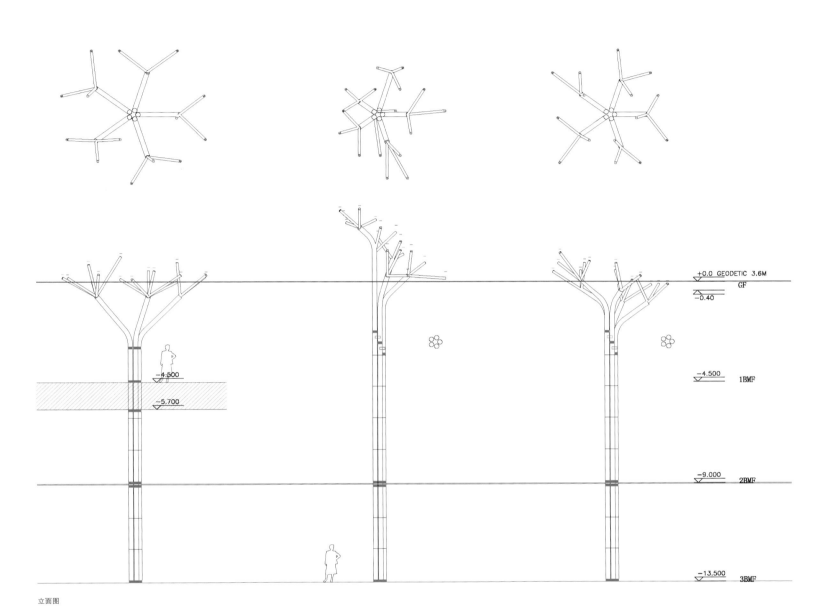

+0.0 GEODETIC 3.6M
GF
-0.40

-4.500

-4.500 1BMF

-5.700

-9.000 2BMF

-13.500 3BMF

立面图

上海K11购物艺术中心
K11 Art Mall, Shanghai

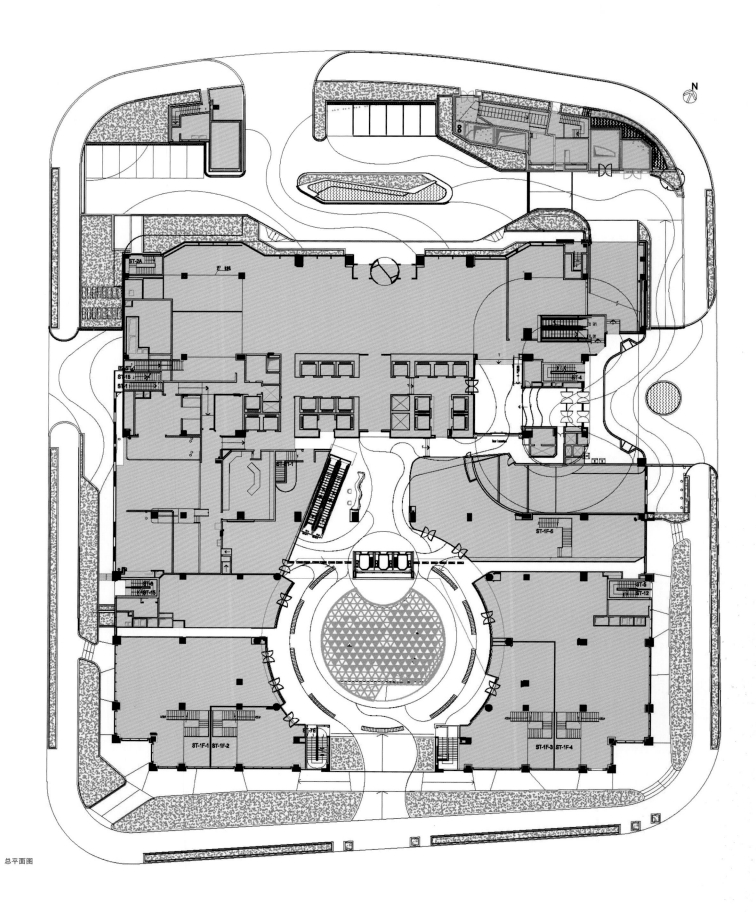

总平面图

地点：上海市　建筑面积：9 100平方米　业主：新世界　设计公司：柯凯建筑设计顾问（上海）有限公司　摄影：Charlie Xia
Location: Shanghai, China　Building Area: 9,100m²　Design Company: Kokaistudios　Photography: Charlie Xia

上海K11购物艺术中心
K11 Art Mall, Shanghai

地点：上海市　建筑面积：9 100平方米　业主：新世界　设计公司：柯凯建筑设计顾问（上海）有限公司　摄影：Charlie Xia
Location: Shanghai, China　Building Area: 9,100m²　Design Company: Kokaistudios　Photography: Charlie Xia

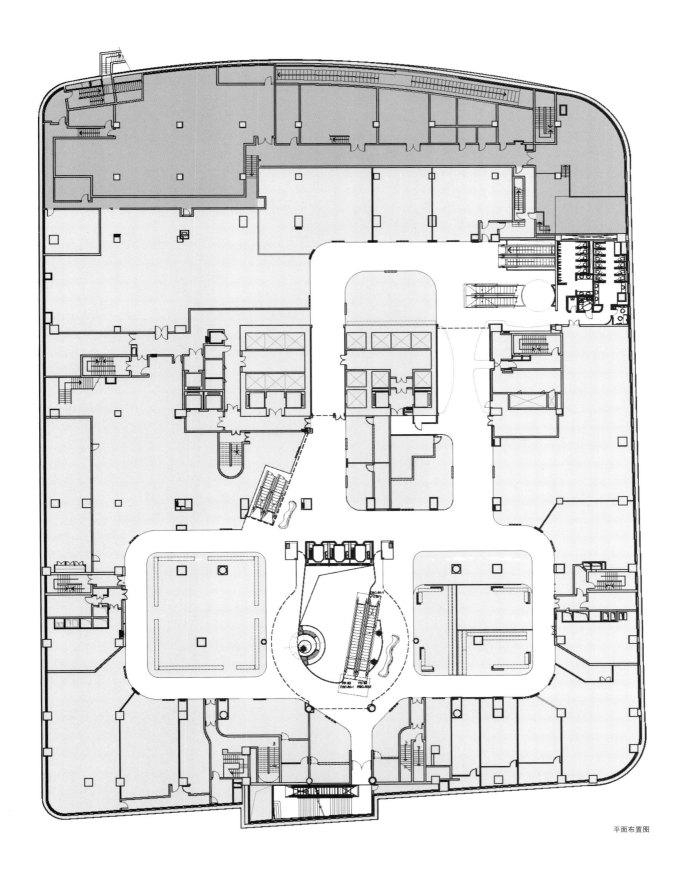

平面布置图

上海K11购物艺术中心
K11 Art Mall, Shanghai

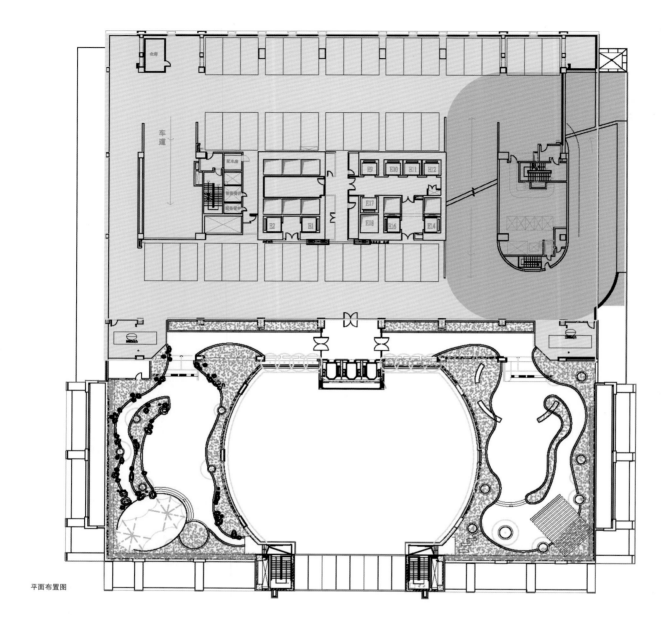

平面布置图

地点: 上海市　建筑面积: 9 100平方米　业主: 新世界　设计公司: 柯凯建筑设计顾问 (上海) 有限公司　摄影: Charlie Xia
Location: Shanghai, China　Building Area: 9,100m²　Design Company: Kokaistudios　Photography: Charlie Xia

上海环茂广场
iAPM

客户：新鸿基地产   建筑设计：英国贝诺建筑师事务所   室内设计：英国贝诺建筑师事务所   总规划设计：英国贝诺建筑师事务所
Client: Sun Hung Kai Properties   Architect: Benoy Ltd   Interior Architect and Designer: Benoy Ltd   Masterplanner: Benoy Ltd

平面造型设计：英国贝诺建筑师事务所　面积（综合部分）：222 000平方米　面积（商业）：110 000平方米　美国绿色建筑认证：金级认证　楼层数：7（地上），2（地下）　商户数：238　项目地点：中国上海淮海路
Size (Mixed-Use Scheme): 222,000m²　Size (Retail) :110,000m²　LEED Award: Gold　Number of Floors: 7 (above ground), 2 (below ground)　Number of Tenants: 238　Location: Huai Hai Road, Shanghai, China

上海环茂广场
iAPM

客户：新鸿基地产　　建筑设计：英国贝诺建筑师事务所　　室内设计：英国贝诺建筑师事务所　　总规划设计：英国贝诺建筑师事务所
Client: Sun Hung Kai Properties　Architect: Benoy Ltd　Interior Architect and Designer: Benoy Ltd　Masterplanner: Benoy Ltd

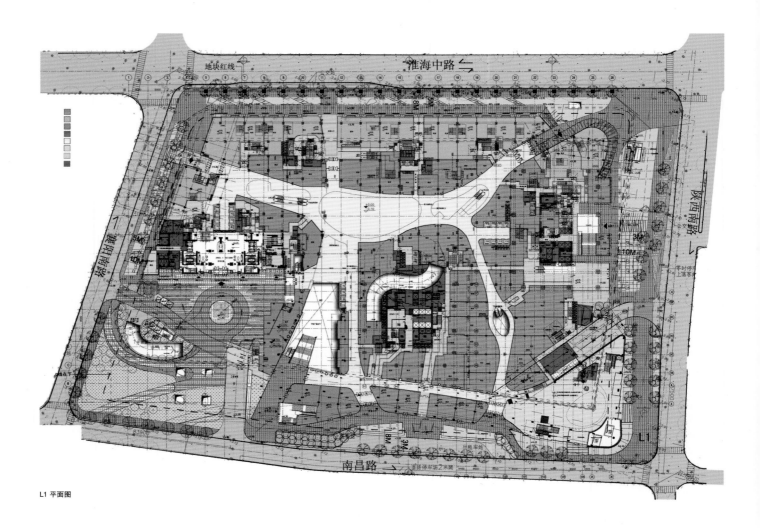

L1 平面图

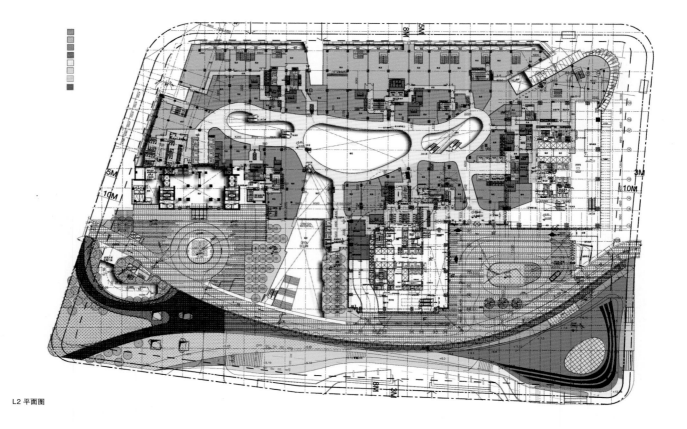

L2 平面图

平面造型设计: 英国贝诺建筑师事务所　面积(综合部分): 222 000平方米　面积(商业): 110 000平方米　美国绿色建筑认证: 金级认证　楼层数: 7(地上), 2(地下)　商户数: 238　项目地点: 中国上海淮海路

Size (Mixed-Use Scheme): 222,000m²　Size (Retail) :110,000m²　LEED Award: Gold　Number of Floors: 7 (above ground), 2 (below ground)　Number of Tenants: 238　Location: Huai Hai Road, Shanghai, China

上海环茂广场
iAPM

客户：新鸿基地产　建筑设计：英国贝诺建筑师事务所　室内设计：英国贝诺建筑师事务所　总规划设计：英国贝诺建筑师事务所
Client: Sun Hung Kai Properties　Architect: Benoy Ltd　Interior Architect and Designer: Benoy Ltd　Masterplanner: Benoy Ltd

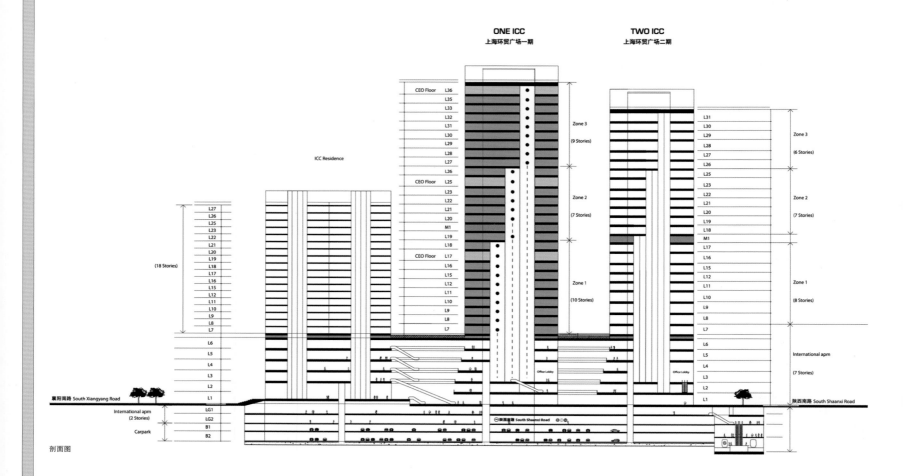

剖面图

平面造型设计: 英国贝诺建筑师事务所　面积(综合部分): 222 000平方米　面积(商业): 110 000平方米　美国绿色建筑认证: 金级认证　楼层数: 7(地上), 2(地下)　商户数: 238　项目地点: 中国上海淮海路

Size (Mixed-Use Scheme): 222,000m²　Size (Retail): 110,000m²　LEED Award: Gold　Number of Floors: 7 (above ground), 2 (below ground)　Number of Tenants: 238　Location: Huai Hai Road, Shanghai, China

上海环茂广场
iAPM

客户：新鸿基地产　建筑设计：英国贝诺建筑师事务所　室内设计：英国贝诺建筑师事务所　总规划设计：英国贝诺建筑师事务所
Client: Sun Hung Kai Properties　Architect: Benoy Ltd　Interior Architect and Designer: Benoy Ltd　Masterplanner: Benoy Ltd

平面造型设计：**英国贝诺建筑师事务所**　面积（综合部分）：**222 000平方米**　面积（商业）：**110 000平方米**　美国绿色建筑认证：**金级认证**　楼层数：**7（地上），2（地下）**　商户数：**238**　项目地点：**中国上海淮海路**
Size (Mixed-Use Scheme): 222,000m² 　Size (Retail):110,000m²　　LEED Award: Gold　Number of Floors: 7 (above ground), 2 (below ground)　Number of Tenants: 238　Location: Huai Hai Road, Shanghai, China

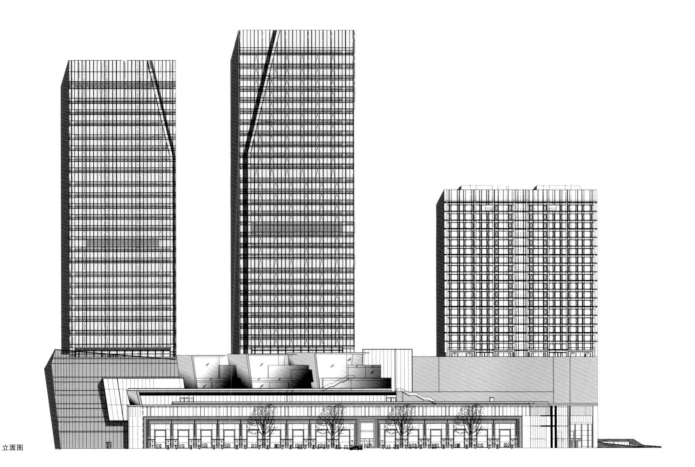

立面图

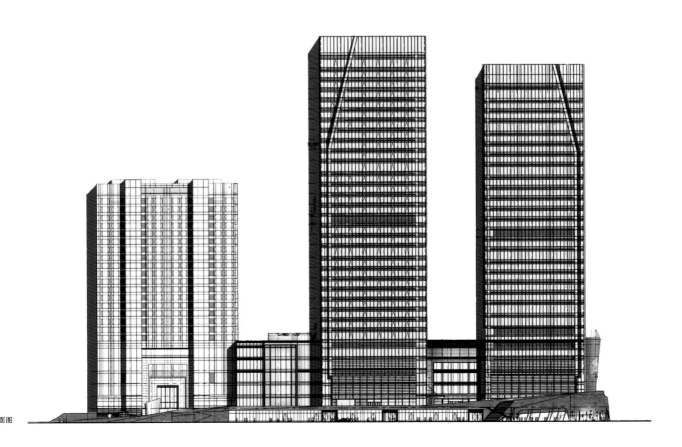

立面图

上海环茂广场
iAPM

客户：新鸿基地产 建筑设计：英国贝诺建筑师事务所 室内设计：英国贝诺建筑师事务所 总规划设计：英国贝诺建筑师事务所
Client: Sun Hung Kai Properties Architect: Benoy Ltd Interior Architect and Designer: Benoy Ltd Masterplanner: Benoy Ltd

上海环茂广场是一个由新型高端商业裙楼组成的综合项目。以两栋A级国际写字塔楼、豪华住宅塔楼及与地铁无缝连接为特征，上海环茂广场代表了这座城市综合商业项目的新未来。紧跟全日无休息经营的理念，潮流商业部分是上海第一个夜间经营的购物广场，同时建立了一个时尚、餐饮及娱乐热点。

这个项目醒目地坐落于淮海路边上海有名的购物区。得益于此区域的法国租界遗址及现代影响，上海环茂广场的设计为这座城市创造了一个新的建筑类型。

裙楼的外立面设计参考了传统的石库门建筑。220米长的醒目的有着不同高度的临街店面使此项目吸引了领先的零售商们，如华伦天奴、普拉达及古琦。

裙楼立面外的林荫大道式的露台模仿下面绿树成行的街道，同时把地面的步行体验转移到上海环茂广场的楼层上。沿着露台，三个现代玻璃"盒子"悬挑出来为餐饮店提供迷人的空间及城市的景色。附加的"下沉"和"冬天"花园为访客提供了宁静的绿色休息地。

室内，位于主入口处创新设计的弹出式商业盒子给商场创造了一个整体性的组合。一面多媒体LED墙向室内空间注入颜色并融入科技。围着一个令人印象深刻的室内中庭及活动空间设计的商业拱廊以双层高的商店、玻璃栏杆及中性色调为特色，补充了商业空间。

这一高档的开发项目充满商业吸引力，拥有超过230个租户，包括新兴的奢侈品零售商、精品餐厅、最先进的影院及生活形态超市——其中60%是国际品牌。

立面图

立面图

平面造型设计：英国贝诺建筑师事务所　面积（综合部分）：222 000平方米　面积（商业）：110 000平方米　美国绿色建筑认证：金级认证　楼层数：7（地上），2（地下）　商户数：238　项目地点：中国上海淮海路
Size (Mixed-Use Scheme): 222,000m²　Size (Retail) :110,000m²　LEED Award: Gold　Number of Floors: 7 (above ground), 2 (below ground)　Number of Tenants: 238　Location: Huai Hai Road, Shanghai, China

上海环茂广场
iAPM

客户：新鸿基地产　建筑设计：英国贝诺建筑师事务所　室内设计：英国贝诺建筑师事务所　总规划设计：英国贝诺建筑师事务所
Client: Sun Hung Kai Properties　Architect: Benoy Ltd　Interior Architect and Designer: Benoy Ltd　Masterplanner: Benoy Ltd

平面造型设计：英国贝诺建筑师事务所　面积（综合部分）：222 000平方米　面积（商业）：110 000平方米　美国绿色建筑认证：金级认证　楼层数：7（地上），2（地下）　商户数：238　项目地点：中国上海淮海路
Size (Mixed-Use Scheme): 222,000m² 　Size (Retail) :110,000m² 　LEED Award: Gold 　Number of Floors: 7 (above ground), 2 (below ground) 　Number of Tenants: 238 　Location: Huai Hai Road, Shanghai, China

上海环茂广场
iAPM

客户：新鸿基地产　建筑设计：英国贝诺建筑师事务所　室内设计：英国贝诺建筑师事务所　总规划设计：英国贝诺建筑师事务所
Client: Sun Hung Kai Properties　Architect: Benoy Ltd　Interior Architect and Designer: Benoy Ltd　Masterplanner: Benoy Ltd

平面造型设计：英国贝诺建筑师事务所　面积（综合部分）：222 000平方米　面积（商业）：110 000平方米　美国绿色建筑认证：金级认证　楼层数：7（地上），2（地下）　商户数：238　项目地点：中国上海淮海路

Size (Mixed-Use Scheme): 222,000m²　Size (Retail) :110,000m²　LEED Award: Gold　Number of Floors: 7 (above ground), 2 (below ground)　Number of Tenants: 238　Location: Huai Hai Road, Shanghai, China

045

上海环茂广场
iAPM

客户: 新鸿基地产   建筑设计: 英国贝诺建筑师事务所   室内设计: 英国贝诺建筑师事务所   总规划设计: 英国贝诺建筑师事务所
Client: Sun Hung Kai Properties   Architect: Benoy Ltd   Interior Architect and Designer: Benoy Ltd   Masterplanner: Benoy Ltd

平面造型设计: **英国贝诺建筑师事务所**   面积(综合部分): **222 000平方米**   面积(商业): **110 000平方米**   美国绿色建筑认证: **金级认证**   楼层数: **7(地上), 2(地下)**   商户数: **238**   项目地点: **中国上海淮海路**
Size (Mixed-Use Scheme): 222,000m²   Size (Retail) :110,000m²   LEED Award: Gold   Number of Floors: 7 (above ground), 2 (below ground)   Number of Tenants: 238   Location: Huai Hai Road, Shanghai, China

iAPM is a new high-end retail podium and the key feature for the Shanghai ICC mixed-use scheme. Featuring two Grade A international office towers, luxury residential tower and seamless connections to the MTR, Shanghai ICC with iAPM represents the new future for mixed-use retail schemes in the city. Following the concept of am to pm, the trendy retail development is the first nighttime shopping mall in Shanghai and has established a hot spot for fashion, dining and entertainment.

The development is prominently located along Huai Hai Road, Shanghai's famous shopping strip. Celebrating the area's French concession heritage as well as contemporary influences, the design for iAPM has created a new architectural typology and icon for the city.

The façade for the podium has taken cues from traditional Shikumen architecture.Producing striking multi-height shop fronts along the rare 220m frontage to the famous street, the commercial benefits have attracted leading retailers such as Valentino, Prada and Gucci to iAPM.

The boulevard-style terraces on the podium façade have mimicked the tree-lined streets below and transported the pedestrian experience to the upper floors of iAPM. Along the terraces, three modern glass 'boxes' jut out and provide intriguing spaces for dining outlets with views across the city. Additional 'Sunken' and 'Winter' gardens provide restful, green havens for visitors.

Inside, innovative pop-out retail boxes at the main entrance create a holistic composition for the mall. A multi-media LED wall injects colour and integrates technology into the interior. Planned around an impressive interior atria and events space, the retail arcades feature double-height shops, glass balustrading and a neutral colour palette to complement the retailers.

The upscale development has been commercially attractive and is fully let with over 230 tenants, including new-to-market and luxury retailers, fine dining restaurants, state-of-the-art cinema and lifestyle supermarket – 60% of which are international brands.

上海环茂广场
iAPM

客户：新鸿基地产　建筑设计：英国贝诺建筑师事务所　室内设计：英国贝诺建筑师事务所　总规划设计：英国贝诺建筑师事务所
Client: Sun Hung Kai Properties　Architect: Benoy Ltd　Interior Architect and Designer: Benoy Ltd　Masterplanner: Benoy Ltd

上海环茂广场
iAPM

客户：新鸿基地产　建筑设计：英国贝诺建筑师事务所　室内设计：英国贝诺建筑师事务所　总规划设计：英国贝诺建筑师事务所

平面造型设计：**英国贝诺建筑师事务所**　面积（综合部分）：**222 000平方米**　面积（商业）：**110 000平方米**　美国绿色建筑认证：**金级认证**　楼层数：**7（地上），2（地下）**　商户数：**238**　项目地点：**中国上海淮海路**
Size (Mixed-Use Scheme): 222,000m²　Size (Retail) :110,000m²　LEED Award: Gold　Number of Floors: 7 (above ground), 2 (below ground)　Number of Tenants: 238　Location: Huai Hai Road, Shanghai, China

上海环茂广场
iAPM

客户：新鸿基地产　建筑设计：英国贝诺建筑师事务所　室内设计：英国贝诺建筑师事务所　总规划设计：英国贝诺建筑师事务所
Client: Sun Hung Kai Properties　Architect: Benoy Ltd　Interior Architect and Designer: Benoy Ltd　Masterplanner: Benoy Ltd

平面造型设计：**英国贝诺建筑师事务所**　面积（综合部分）：**222 000平方米**　面积（商业）：**110 000平方米**　美国绿色建筑认证：**金级认证**　楼层数：**7（地上），2（地下）**　商户数：**238**　项目地点：**中国上海淮海路**
Size (Mixed-Use Scheme): 222,000m²　Size (Retail) :110,000m²　LEED Award: Gold　Number of Floors: 7 (above ground), 2 (below ground)　Number of Tenants: 238　Location: Huai Hai Road, Shanghai, China

哈尔滨哈西万达广场
Haxi Wanda Plaza, Harbin

地点：黑龙江省哈尔滨市西新区　占地面积：186 500平方米　建筑面积：863 500平方米　开发商：万达集团　立面设计：鼎实国际
Location: Xixin District, Harbin, Heilongjiang Province, China　Site Area: 186,500m²　Building Area: 863,500m²　Developer: Wanda Group　Facade Design: De-sign Architectural

哈尔滨哈西万达广场
Haxi Wanda Plaza, Harbin

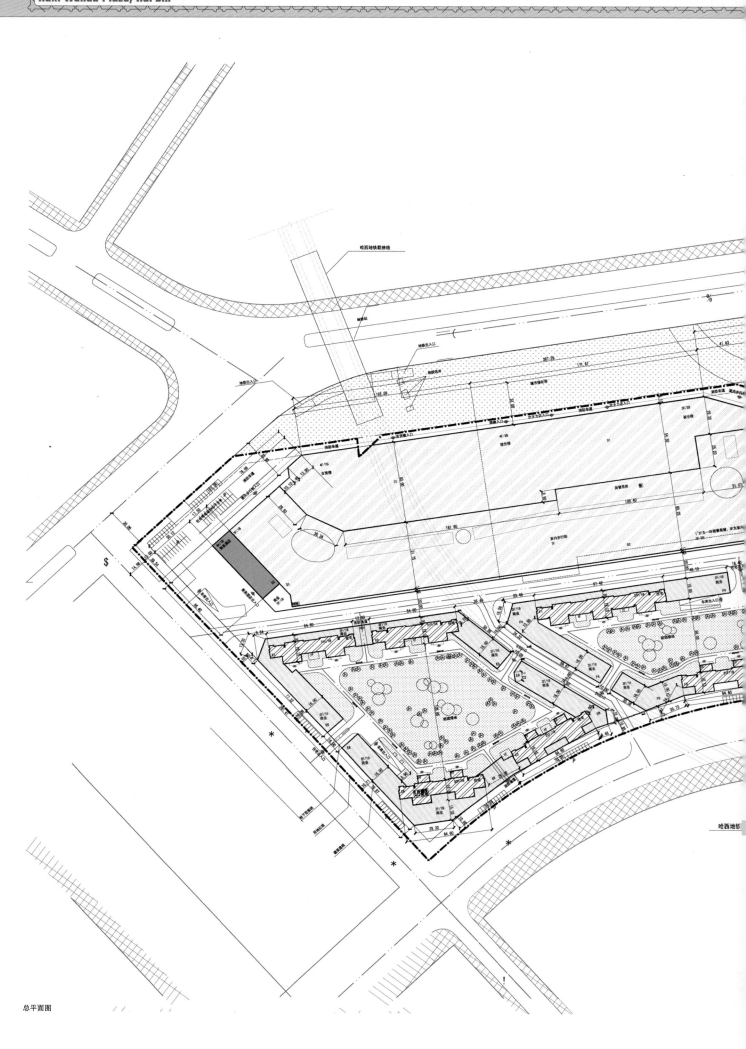

总平面图

地点：黑龙江省哈尔滨市西新区　占地面积：186 500平方米　建筑面积：863 500平方米　开发商：万达集团　立面设计：鼎实国际
Location: Xixin District, Harbin, Heilongjiang Province, China　Site Area: 186,500m²　Building Area: 863,500m²　Developer: Wanda Group　Facade Design: De-sign Architectural

哈尔滨哈西万达广场
Haxi Wanda Plaza, Harbin

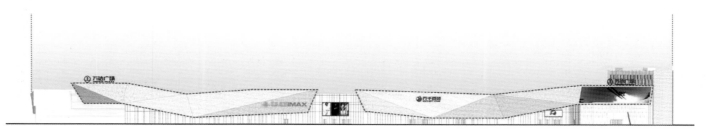

沿中兴大道底层商业以玻璃橱窗不主，创造通透的视觉效果。

商业主入口采用菱形玻璃幕墙，营造浓厚的商业氛围。

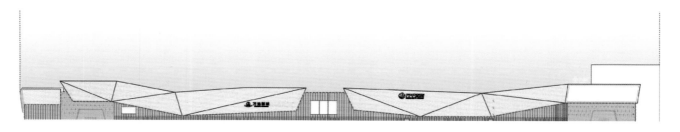

体块 Massing   表皮 Skin   肌理 Texture

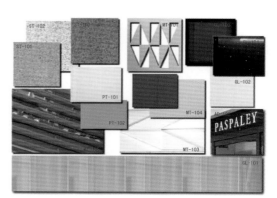

立面图及节点

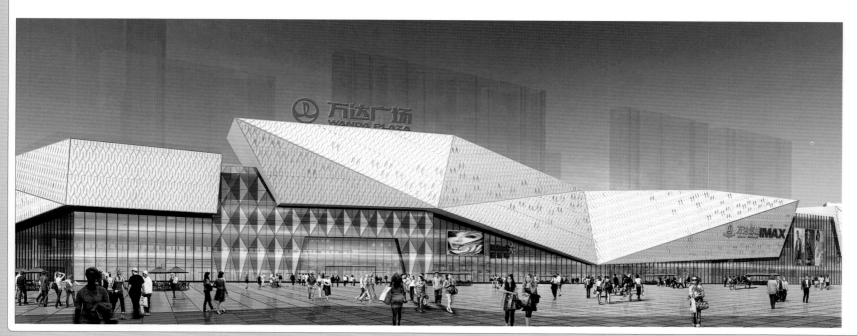

地点：黑龙江省哈尔滨市西新区　占地面积：186 500平方米　建筑面积：863 500平方米　开发商：万达集团　立面设计：鼎实国际
Location: Xixin District, Harbin, Heilongjiang Province, China　Site Area: 186,500m²　Building Area: 863,500m²　Developer: Wanda Group　Facade Design: De-sign Architectural

哈尔滨哈西万达广场项目是万达集团斥巨资打造建设的哈尔滨第三座万达广场，是目前东三省规模最大的城市综合体。哈西万达广场集大型购物中心、量贩式KTV、电玩城、国际影城、国际连锁超市、星级酒店、室外步行商业街、高档写字楼、精装SOHO公寓及高档住宅于一体，总规划用地186 500平方米，建筑面积863 500平方米。其中大商业面积210 000平方米，写字楼180 000平方米，五星级酒店40 000平方米，住宅68 000平方米，建成后成为该区域的新城市地标。

本次设计从所处的当地文化特色出发，将哈尔滨"冰城"的概念引入设计中，通过不同的手法将此概念表现在立面体块、表面肌理、入口玻璃幕墙等位置，既丰富了商业立面，使建筑富有时代气息，又与当地文化交相辉映。

本设计的立面造型概念取自海上浮冰瑰丽壮阔的形体，本着文化与建筑立面完美结合的设计理念，立面设计充分发掘城市文脉，结合冰城的地域特色，将商业主体设计成梦幻的冰块，用现代建筑手段勾勒出冰纹的脉络，转折的线条犹如舞者在冰上舞蹈。整个广场的形体简洁、现代、大气并气势恢宏，宛如一尊巨大的冰雪世界艺术品，展现出一个造型独特带有浓郁地方特色和时尚气息的商业文化中心。

整个广场的表皮肌理呈现出雪花的自然纹理，并向建筑表面及室外景观延续，既突出了哈尔滨当地的特色，又借助现代的设计手法。设计师使用白色铝板结合点状玻璃的形式，将雪花飞舞的缤纷造型表现在建筑立面上，突出其超现实主义、抽象灵动、仿生自然的特色。

在贯彻万达广场既有风格的基础上，通过对重点立面节点的处理来实现对主要方向城市人流的引导，通过入口处理，来激活主要城市界面的商业氛围。采用体块转折的方式来弱化扁长立面的单调感，在两侧入口部位以菱形玻璃为构成元素并结合上方挑出"盒子"造型的广告屏幕，形成更为丰富的商业空间。

按着万达广场一贯的标志系统设计原则，在塔楼顶部及商业步行街的入口顶部设置了万达企业的LOGO。大商业形体右侧被设计成挑出的广告灯箱，与下方的入口完美结合，既突出入口，又加强对人流的引导性。

酒店总体轮廓呈简洁流畅的竖向线条，高大挺拔，外立面采用天然石材。线角的丰富细腻，使建筑在转折、进退中增强了动感和光影的和谐变化。

甲级写字楼外立面由石材与玻璃幕墙相互咬合构成，尊贵、时尚的立面元素营造出Art Deco风格的商业氛围。SOHO公寓外立面采用窗墙的做法，材料用真石漆喷涂构成，既降低了造价，又能与写字楼风格统一。

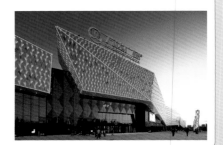

哈尔滨哈西万达广场
Haxi Wanda Plaza, Harbin

地点：黑龙江省哈尔滨市西新区　占地面积：186 500平方米　建筑面积：863 500平方米　开发商：万达集团　立面设计：鼎实国际
Location: Xixin District, Harbin, Heilongjiang Province, China　Site Area: 186,500m²　Building Area: 863,500m²　Developer: Wanda Group　Facade Design: De-sign Architectural

哈尔滨哈西万达广场
Haxi Wanda Plaza, Harbin

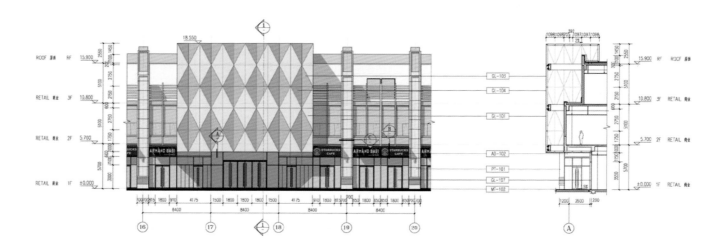

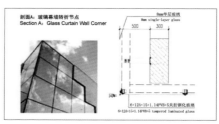

剖面A：玻璃幕墙转折节点
Section A: Glass Curtain Wall Corner

6+12A+15+1.14PVB+5夹胶钢化玻璃
6+12A+15+1.14PVB+5 tempered laminated glass

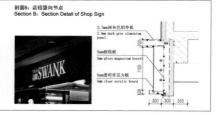

剖面B：店铺招牌竖向节点
Section B: Section Detail of Shop Sign

2.5mm深灰色铝单板
2.5mm dark grey aluminium panel

5mm玻镁板
5mm glass magnesium board

5mm透明亚克力板
5mm clear acrylic board

剖面C：壁柱横剖节点
Section C: Cross Section of Pilaster

外墙仿石材涂料(耐-40)
exterior wall stone-like paint

大商业南立面示意图
South Elevation of
Shopping Mall

玻璃幕墙意向图
Reference for Glass Curtain Wall

店铺招牌意向图
Reference for Shop Sign

金属格栅意向图
Reference for Metal Grille

立面图及节点

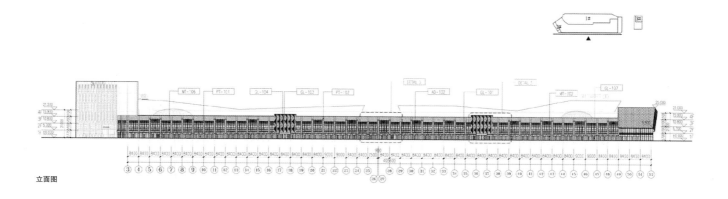

立面图

地点：黑龙江省哈尔滨市西新区　占地面积：186 500平方米　建筑面积：863 500平方米　开发商：万达集团　立面设计：鼎实国际
Location: Xixin District, Harbin, Heilongjiang Province, China　Site Area: 186,500m²　Building Area: 863,500m²　Developer: Wanda Group　Facade Design: De-sign Architectural

立面图及节点图

广告牌意向图
Reference for Advertising Board

夜景照明效果图
Rendering -Lighting at Night

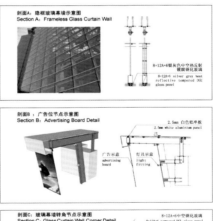

剖面A：隐框玻璃幕墙示意图
Section A：Frameless Glass Curtain Wall

8+12A+6银灰色中空热反射镀膜钢化玻璃
8+12A+6 silver grey heat reflective tempered DGL glass panel

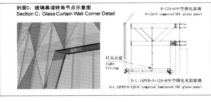

剖面B：广告位节点示意图
Section B：Advertising Board Detail

2.5mm 白色铝单板
2.5mm white aluminium panel

广告牌
advertising board

灯具示意
light fitting

剖面C：玻璃幕墙转角节点示意图
Section C：Glass Curtain Wall Corner Detail

8+12A+6中空钢化玻璃
8+12A+6 tempered DGL glass panel

灯具示意
light fitting

5+1.14PVB+5+12A+6中空钢化夹胶玻璃
6+1.18PVB+5+12A+6 tempered laminated DGL glass panel

GL-104　深蓝色镀膜LOW-E中空玻璃
Dark Blue LOW-E Reflective DGU Glass Panel

MT-103　白色铝板
White Aluminium Panel

北立面示意图
North Elevation

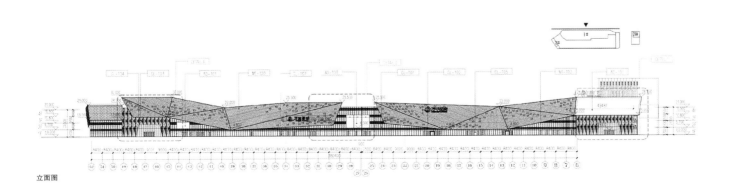

立面图

哈尔滨哈西万达广场
Haxi Wanda Plaza, Harbin

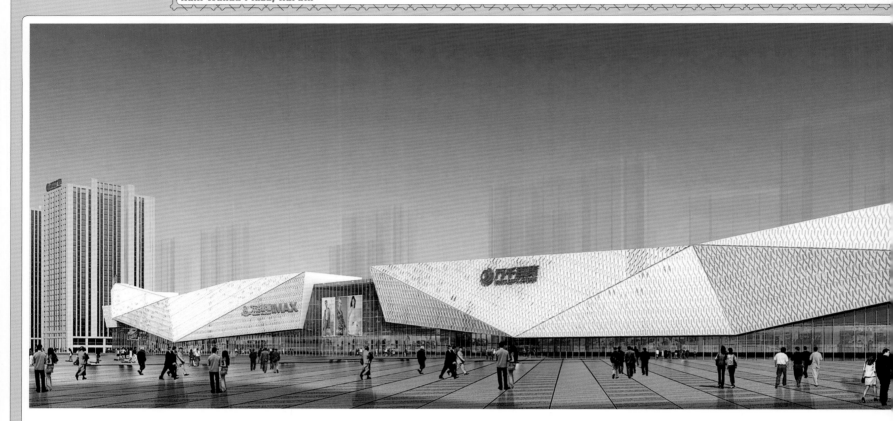

The project is the third Wanda Plaza in Harbin built by Wanda Group, and it is the largest urban complex across the three northeast provinces of China currently. The project gathers large shopping center, KTV, game center, international cinema, international chain supermarket, star hotel, indoor and outdoor commercial walking street, high-end office building, hardcover SOHO apartment and high-class residence, with site area of 186,500m^2

and building area of 863,500m^2, of which, commercial area occupies 210,000m^2, office area of 180,000m^2, five-star hotel area of 40,000m^2 and residence area of 68,000m^2. After completion, the project becomes a new city landmark.

Starting from the local culture, the design draws in the concept of "ice city". Through varied approaches, the concept is embodied on the facade mass, surface texture, glass curtain wall

at the entrance, etc., which not only enriches the commercial facade and endows the building with time breath, but also echoes local culture.

The facade shape is inspired from the grand shape of sea ice. To realize a perfect combination of local culture and building facade, the design digs deep into local culture and draws in the concept of "ice city", thus designs the commercial body into a dreamy ice block. Ice texture is outlined through

modern architectural method, and the winding lines are just like dancers dancing on the ice. The project is like a huge ice art filled with local culture and modern breath.

The natural vein of snowflake is repeated on the building facade and exterior landscape, which not only highlights the local feature of Harbin, but also reveals features of surrealism, abstraction, flexibility and nature simulation through modern design

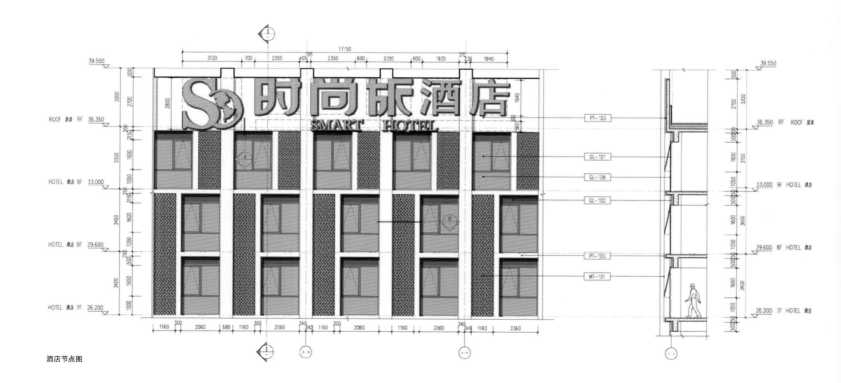

酒店节点图

地点：黑龙江省哈尔滨市西新区　占地面积：186 500平方米　建筑面积：863 500平方米　开发商：万达集团　立面设计：鼎实国际
Location: Xixin District, Harbin, Heilongjiang Province, China　Site Area: 186,500m²　Building Area: 863,500m²　Developer: Wanda Group　Facade Design: De-sign Architectural

method, the combination of white aluminum plate and point-pattern glass, and presence of dancing snowflake on facade.

On the premise of following the existing style of Wanda Plazas, the design realizes guiding customer flow through important facade node design, and activates the commercial atmosphere of main urban surface through entrance design. The bending of volume weakens the monotonous feel caused by the prolate facade. At the entrances of the two sides, rhombus glass serves as the component element and forms richer commercial space when combined with the upper jetting-out box-shape ad screen.

According to the consistent sign system of Wanda Plaza, the logo is placed on the top of towers, as well as the entrance top of commercial walking street. The right side of the commercial body is designed as a jetting-out ad lamp box, echoing and highlighting the entrance below, and guiding customer flow more effectively.

The hotel silhouette features concise and smooth vertical lines, with facade covered by natural stone. The rich and delicate lines strengthen dynamic sense and the alternation of light and shadow.

The office building combines stone and glass on its facade. These noble yet fashionable facade elements create an Art Deco commercial ambiance. The facade of SOHO apartment adopts real stone paint, which not only reduces cost, but also echoes the style of office building.

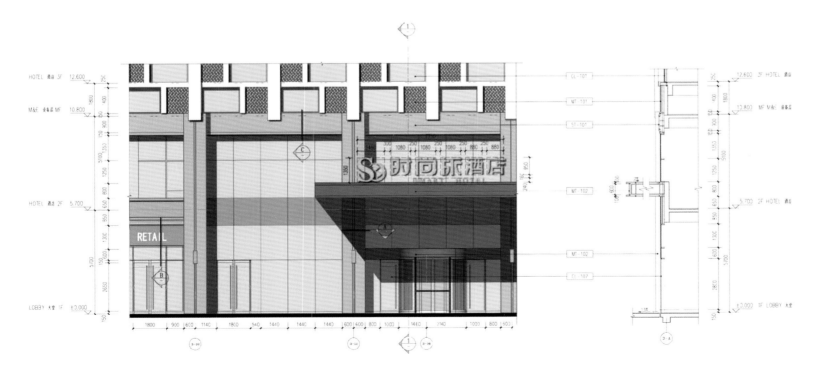

哈尔滨哈西万达广场
Haxi Wanda Plaza, Harbin

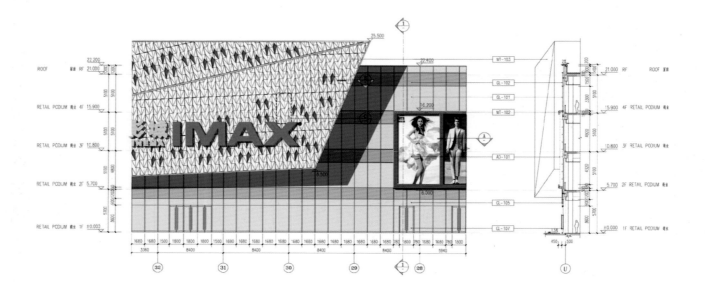

---

1 1#节点2 立面详图　　　1:300
Detail 2　Elevation　　　Scale 1:300

2 1#节点2 1-1剖面图　　　1:300
Detail 2　Section 1-1　　Scale 1:300

剖面A: 广告位铝板包边节点示意图
Section A: Detail of Advertising Board

MT-103 白色造型铝板
3D White Aluminium Panel

剖面B: 铝板幕墙灯槽示意图
Section B: Light Channels on Aluminium Cladding

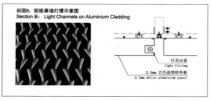

GL-101 LOW-E中空玻璃
Low-E DGU Glass Panel

广告牌夜景照明意向图
Reference-Advertising Board Lighting / Night Scene

铝板幕墙夜景照明意向图
Reference-Aluminum Cladding Lighting / Night Scene

剖面C: 造型铝板大样图
Section C: Aluminium Panel Detail

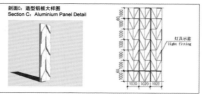

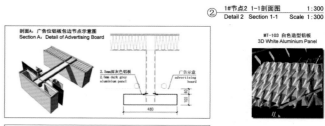

立面图及节点

北立面示意图
North Elevation

剖面A: 立面石材构件示意图
Section A: Facade Stone Cladding Detail

ST-102 黄金麻石材
Golden Grain Stone

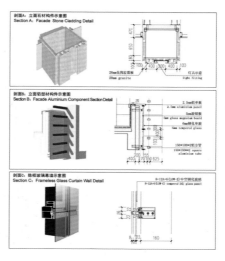

剖面B: 立面铝材构件示意图
Section B: Facade Aluminium Component Section Detail

GL-102 双层中空玻璃
Double Glazed
Insulated Glass

米黄色石材意向图
Reference for Beige Stone

夜景照明效果图
Rendering of Lighting Night Scene

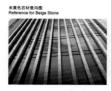

剖面C: 隐框玻璃幕墙示意图
Section C: Frameless Glass Curtain Wall Detail

MT-105 咖啡色铝构件
Brown Aluminium
Component

酒店节点

五星酒店东立面示意图
East Elevation of five-star Hotel

地点：黑龙江省哈尔滨市西新区　占地面积：186 500平方米　建筑面积：863 500平方米　开发商：万达集团　立面设计：鼎实国际
Location: Xixin District, Harbin, Heilongjiang Province, China　Site Area: 186,500m²　Building Area: 863,500m²　Developer: Wanda Group　Facade Design: De-sign Architectural

哈尔滨哈西万达广场
Haxi Wanda Plaza, Harbin

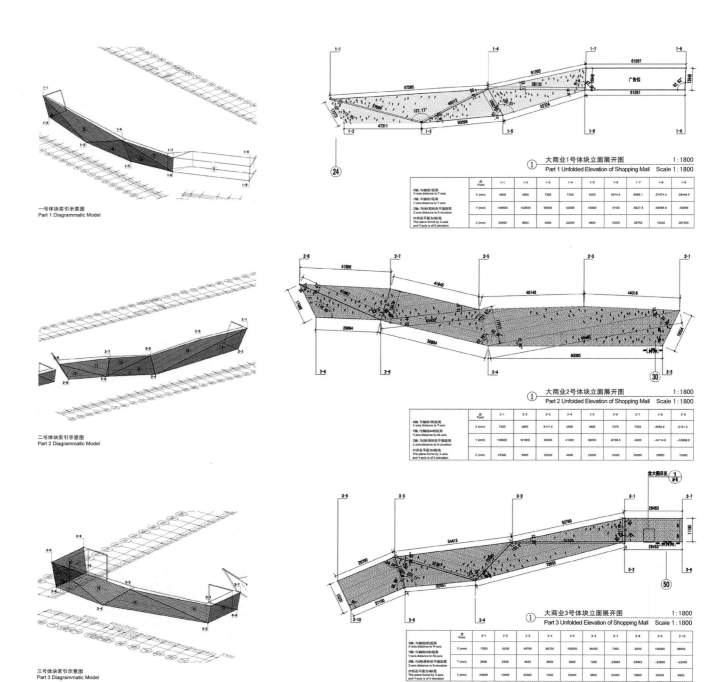

一号体块索引示意图
Part 1 Diagrammatic Model

① 大商业1号体块立面展开图　　1:1800
Part 1 Unfolded Elevation of Shopping Mall　Scale 1:1800

点 Point		1-1	1-2	1-3	1-4	1-5	1-6	1-7	1-8
X (mm)	X轴距：与轴线1的距离 X-axis distance to 1-axis	4800	4800	7300	7300	4200	5914.6	6589.1	-27474.4	-28448.9
Y (mm)	Y轴距：与轴线I的距离 Y-axis distance to I-axis	149600	142600	95600	52600	45600	-6160	-8227.5	-58996.8	-56990
Z (mm)	Z轴距：与0标高在平面距离 Z-axis distance to 0-elevation	20500	9600	4800	22000	4800	15000	28750	15000	287500

XY所在平面为0标高
The plane formd by X-axis
and Y-axis is of 0 elevation

二号体块索引示意图
Part 2 Diagrammatic Model

① 大商业2号体块立面展开图　　1:1800
Part 2 Unfolded Elevation of Shopping Mall　Scale 1:1800

点 Point		2-1	2-2	2-3	2-4	2-5	2-6	2-7	2-8	2-9
X (mm)	X轴距：与轴线1的距离 X-axis distance to 1-axis	7300	4800	6117.5	4800	4800	7070	7300	-5852.6	-5131.2
Y (mm)	Y轴距：与轴线44的距离 Y-axis distance to 44-axis	128000	121500	54000	41000	36000	-8799.5	-5000	-44114.8	-35868.8
Z (mm)	Z轴距：与0标高在平面距离 Z-axis distance to 0-elevation	20500	8500	25000	4800	22000	15000	30000	30000	15000

XY所在平面为0标高
The plane formd by X-axis
and Y-axis is of 0 elevation

三号体块索引示意图
Part 3 Diagrammatic Model

① 大商业3号体块立面展开图　　1:1800
Part 3 Unfolded Elevation of Shopping Mall　Scale 1:1800

点 Point		3-1	3-2	3-3	3-4	3-5	3-6	3-7	3-8	3-9	3-10
X (mm)	X轴距：与轴线R的距离 X-axis distance to R-axis	-7000	-5000	45700	65700	100000	98000	7000	-5000	100000	98000
Y (mm)	Y轴距：与轴线52的距离 Y-axis distance to 52-axis	2500	2500	4600	2600	3800	1200	-23963	-23963	-25000	-25000
Z (mm)	Z轴距：与0标高在平面距离 Z-axis distance to 0-elevation	24000	13000	22000	7000	25000	9600	24000	13000	25000	9600

XY所在平面为0标高
The plane formd by X-axis
and Y-axis is of 0 elevation

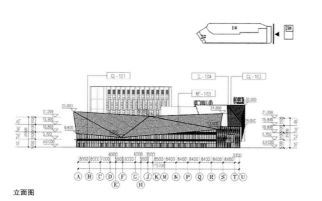

立面图

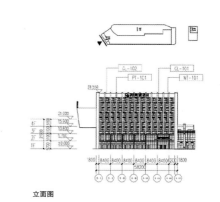

立面图

地点：黑龙江省哈尔滨市西新区　占地面积：186 500平方米　建筑面积：863 500平方米　开发商：万达集团　立面设计：鼎实国际
Location: Xixin District, Harbin, Heilongjiang Province, China　Site Area: 186,500m²　Building Area: 863,500m²　Developer: Wanda Group　Facade Design: De-sign Architectural

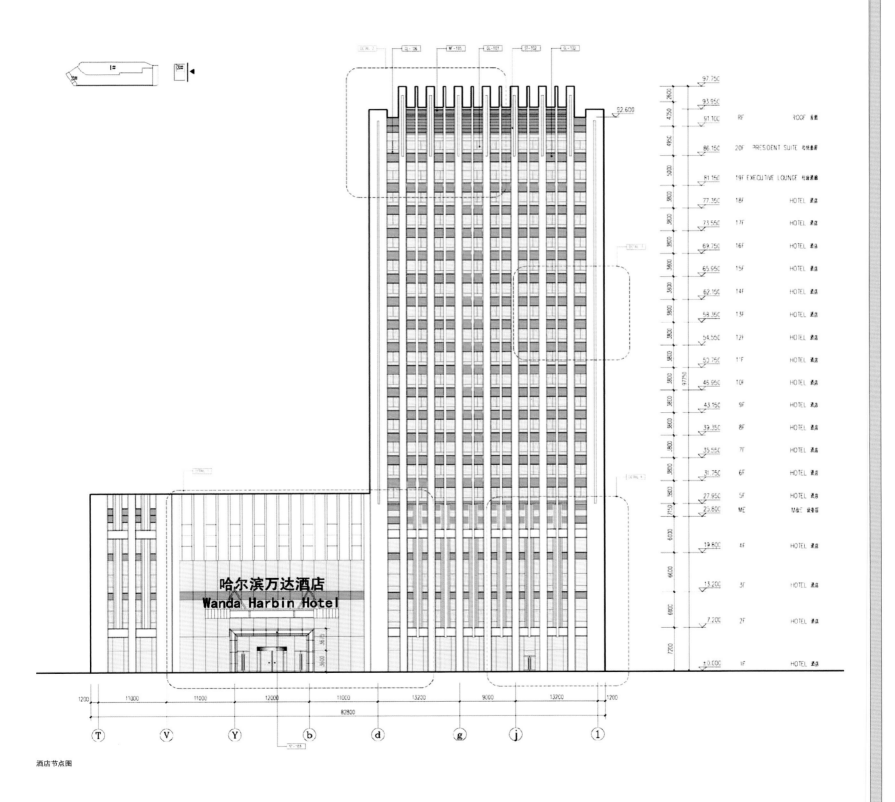

哈尔滨万达酒店
Wanda Harbin Hotel

97.750		
93.950		
91.100	RF	ROOF 屋面
86.150	20F	PRESIDENT SUITE 总统套房
81.150	19F	EXECUTIVE LOUNGE 行政酒廊
77.350	18F	HOTEL 酒店
73.550	17F	HOTEL 酒店
69.750	16F	HOTEL 酒店
65.950	15F	HOTEL 酒店
62.150	14F	HOTEL 酒店
58.350	13F	HOTEL 酒店
54.550	12F	HOTEL 酒店
50.750	11F	HOTEL 酒店
46.950	10F	HOTEL 酒店
43.150	9F	HOTEL 酒店
39.350	8F	HOTEL 酒店
35.550	7F	HOTEL 酒店
31.750	6F	HOTEL 酒店
27.950	5F	HOTEL 酒店
23.800	ME	M&E 设备层
19.800	4F	HOTEL 酒店
13.200	3F	HOTEL 酒店
7.200	2F	HOTEL 酒店
±0.000	1F	HOTEL 酒店

酒店节点图

哈尔滨哈西万达广场
Haxi Wanda Plaza, Harbin

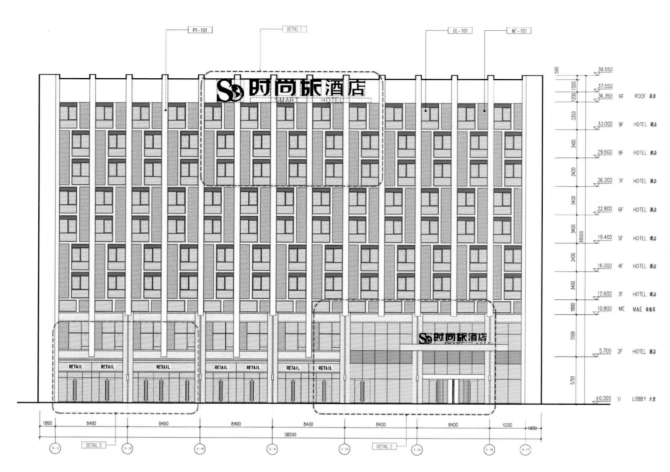

立面图

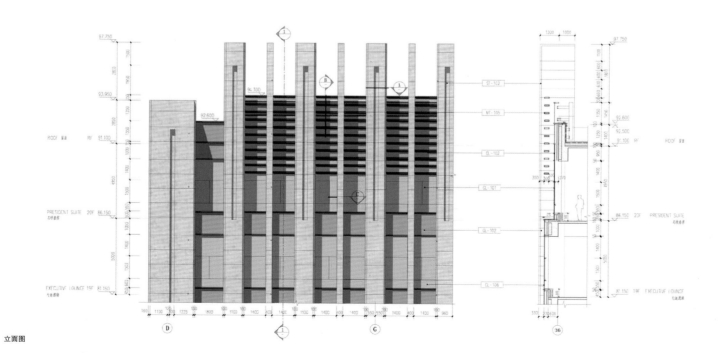

立面图

地点：黑龙江省哈尔滨市西新区　占地面积：186 500平方米　建筑面积：863 500平方米　开发商：万达集团　立面设计：鼎实国际
Location: Xixin District, Harbin, Heilongjiang Province, China　Site Area: 186,500m²　Building Area: 863,500m²　Developer: Wanda Group　Facade Design: De-sign Architectural

LOGO示意图
Reference for LOGO

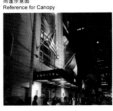

雨蓬示意图
Reference for Canopy

夜景照明示意图
Reference for Lighting / Night Scene

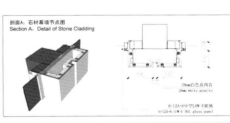

剖面A：石材幕墙节点图
Section A：Detail of Stone Cladding

剖面B：广告位节点图
Section B：Detail of Advertising Board

剖面C：石材幕墙节点图
Section C：Detail of Stone Cladding

ST-101　浅灰色石材
Light Grey Stone

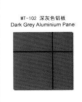

MT-102　深灰色铝板
Dark Grey Aluminium Pane

商务酒店正立面示意图
Front Elevation
of Business Hotel

立面图

抚顺万达广场
Wanda Plaza, Fushun

抚顺万达广场
Wanda Plaza, Fushun

抚顺万达广场
Wanda Plaza, Fushun

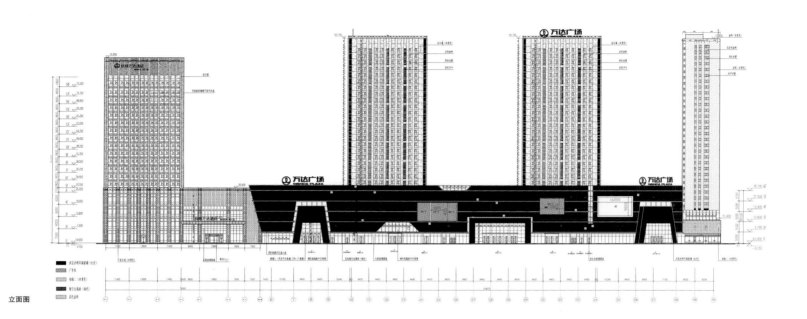

立面图

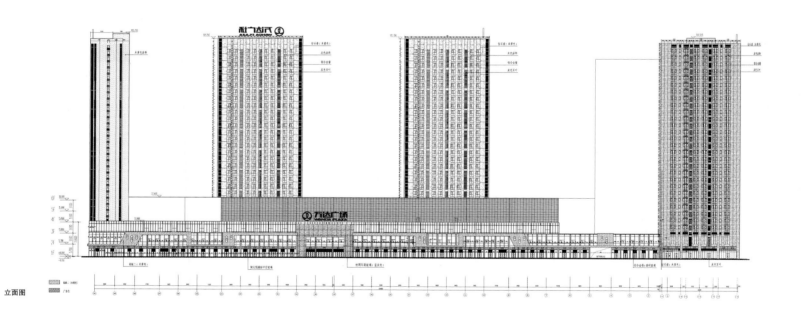

立面图

抚顺万达广场项目位于辽宁抚顺市新抚区，处于抚顺市中心位置，总规划用地面积138 800平方米，总建筑面积932 100平方米。地上建筑面积752 000平方米，其中，购物中心105 700平方米、酒店32 300平方米、写字楼136 400平方米、住宅255 900平方米、商铺61 500平方米、城市配套物业6 500平方米、回迁物业160 500平方米，地下建筑面积180 100平方米。抚顺万达广场的开业填补了抚顺市大型城市综合体的空白。

裙楼外立面设计提取"中式窗花"元素，用镂空金属板以模化块呈现，融合古典建筑文化，试图使用现代的材料和手法与传统的风格相呼应。清朝文化、矿产是抚顺这一城市的特点，设计从中汲取灵感，采用金属板材料，并精心设计了模块化的窗花纹理，达到古典与现代完美结合的效果。镂空金属板形成投影，在阳光的照射下产生美感。

在大商业的主入口设计上，以抚顺盛产的稀有矿物"血珀"为设计主题，在保持了整体感和大气感的同时，利用丝网印刷玻璃形成大体量的门面，吸引人流进入商场，成为强烈的视觉中心。尺度上改变原有的比较冷漠与单调的界面，增强材料质感与种类的统一性，使城市人

群在万达广场内的空间体验得到改善。

抚顺原为大清龙脉所在，采光顶以"龙行天下"为主题，室内设计元素以龙麟为根基，整条室内步行街宛如龙戏二珠。夕阳西下的暮色之中，抚顺万达广场像一条卧龙般潜伏于浑江河畔。

椭圆形中庭是整个空间的重中之重，是整个设计的突出部分。中庭侧板整个形体好似云雾中腾飞的蛟龙，若隐若现。采光顶钢结构和

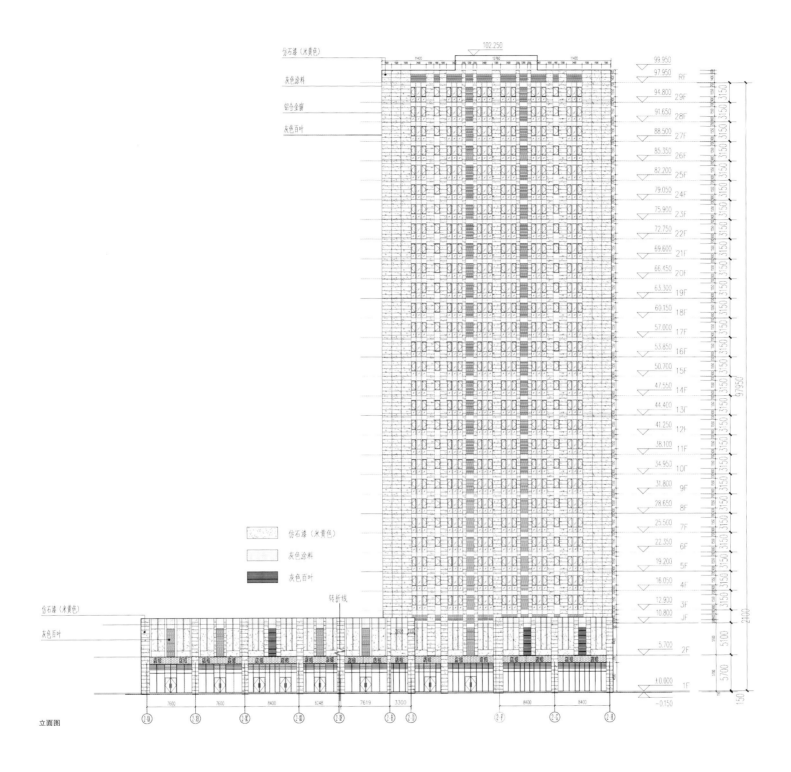

立面图

观光梯的菱形也体现了龙鳞元素，丰富了整个空间。圆形中庭与椭圆形中庭采取了类似的设计手法，但各有不同。圆形中庭两边的侧板以龙鳞花纹为肌理，富有雕塑感，在灯光的作用下，就好像一条腾飞的龙浑身散发着光芒，让人挪不开眼。两边走廊的顶棚造型不再是柔软有弧度的线条，而转变成有棱角的刚硬线条，与椭圆形中庭的顶棚相对应，观光电梯一

改往常用的龙鳞元素，采用方形错落有致地排列——其实还是龙鳞元素演变而来的。
酒店外立面设计采用大气端庄、简洁明快的手法。重视与城市环境的互动关系，使建筑与环境充分协调又独具个性。整体立面吸取了在庄重的古典建筑中常用的三段式设计手法。塔楼顶部采用密集壁柱形式，与主体相辅相成，形成方案的视觉中心。强烈的虚实对比，既传统

又时尚。塔楼中段采用简洁的米黄色仿石涂料加深窗洞的形式，增加了光影变化，立面整体感强，简洁而有力度。裙楼采用四层通高的落地大窗与厚重石材包裹，简洁大气，所有幕墙玻璃分隔挺立，均与塔楼窗洞及玻璃分格对应，"逻辑关系"清晰，对位严谨，同时与入口巨型雨棚呼应，彰显奢华和尊贵的品质。
位于一层的酒店大堂吊顶上空直径10米的大

水晶灯，以抚顺市市花玫瑰为主题。璀璨的水晶勾勒出绚丽的花瓣，尽显奢华大气，主背景墙将从龙袍提取的图案元素加以艺术处理，从色彩纹样上体现浓郁的清朝文化，风格上既延续了万达酒店的传统奢华路线，又将当地人文特征融入其中。

抚顺万达广场
Wanda Plaza, Fushun

抚顺万达广场
Wanda Plaza, Fushun

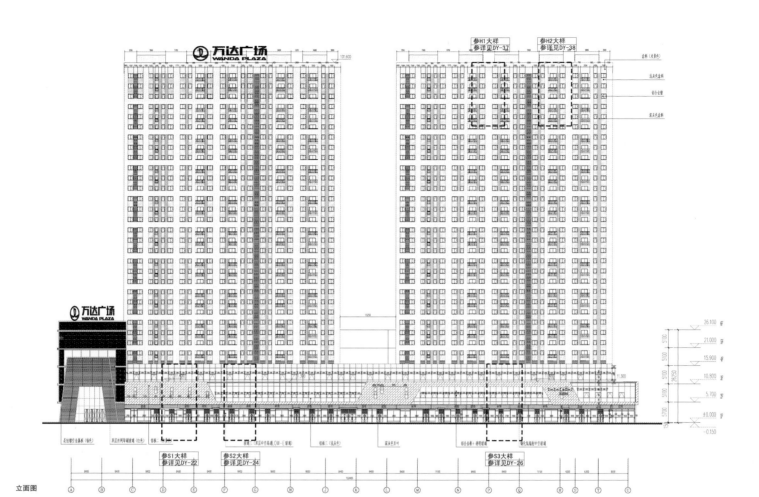

立面图

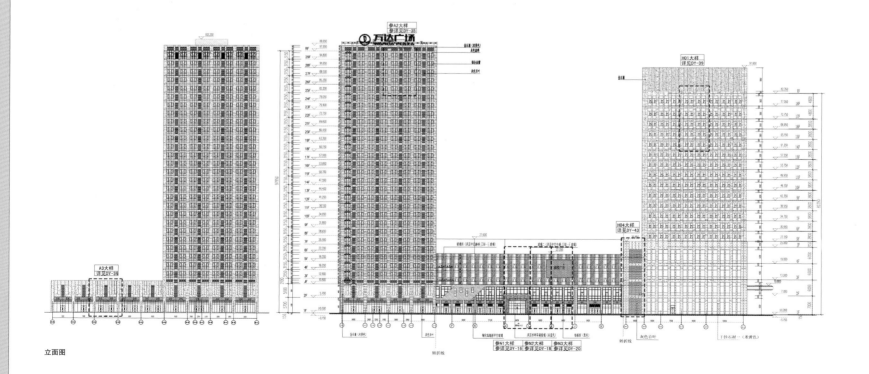

立面图

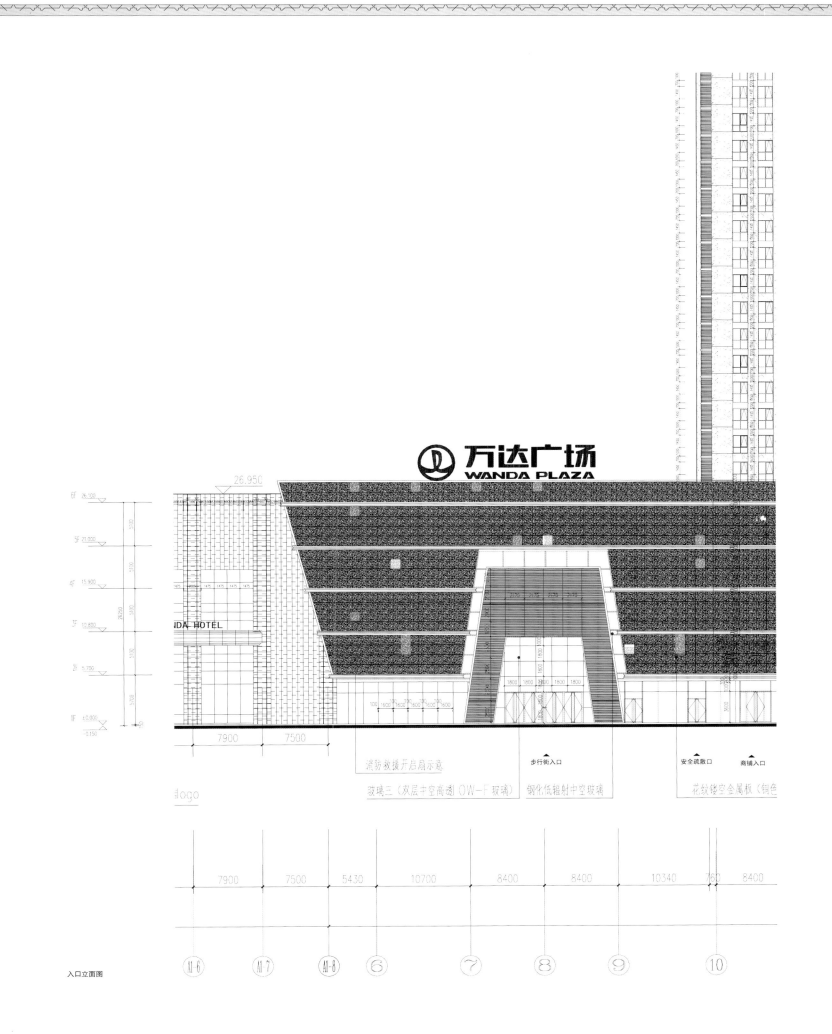

万达广场
WANDA PLAZA

NDA HOTEL

入口立面图

抚顺万达广场
Wanda Plaza, Fushun

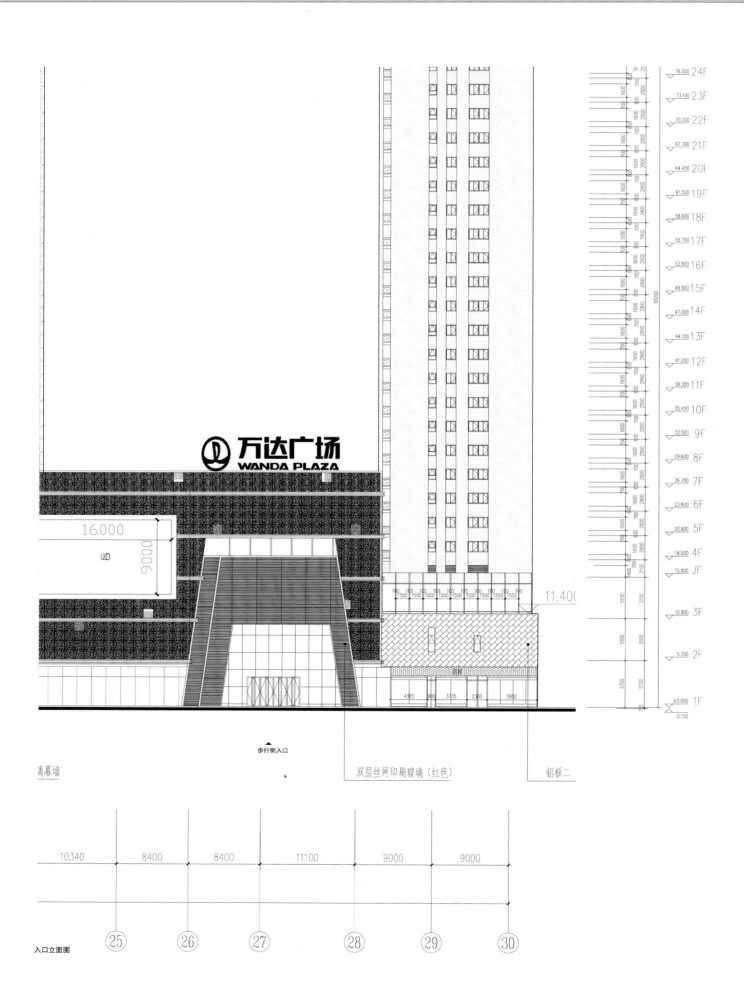

离幕墙 　　　　　双层丝网印刷玻璃（红色）　　　铝板二

步行街入口

入口立面图

地点：辽宁省抚顺市
Location: Fushun, Liaoning Province, China

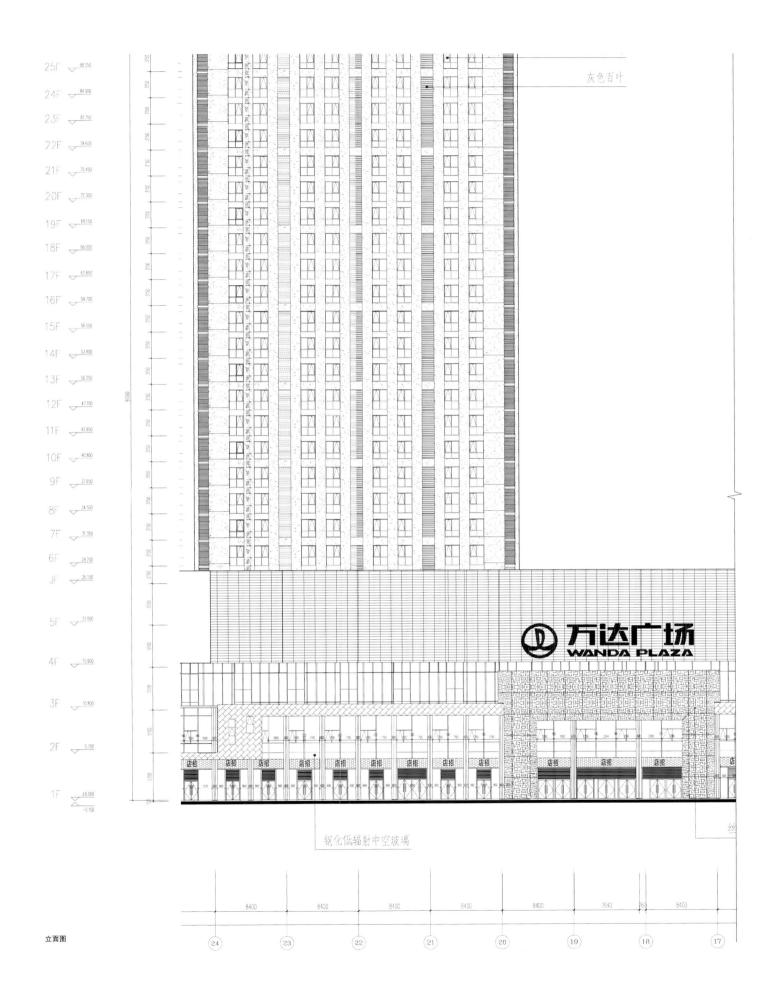

灰色百叶

万达广场
WANDA PLAZA

店招 店招 店招 店招 店招 店招 店招 店招 店招 店招 店招 店招 店招 店招 店招

钢化低辐射中空玻璃

立面图

24 23 22 21 20 19 18 17

25F 24F 23F 22F 21F 20F 19F 18F 17F 16F 15F 14F 13F 12F 11F 10F 9F 8F 7F 6F JF 5F 4F 3F 2F 1F

8400 8400 8400 8400 8400 7640 760 8400

抚顺万达广场
Wanda Plaza, Fushun

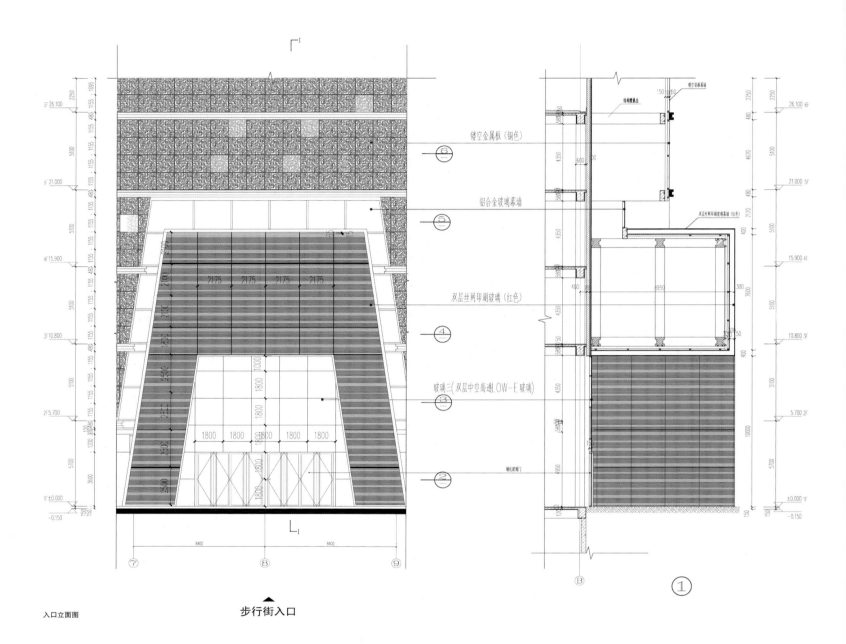

镂空金属板 (铜色)

铝合金玻璃幕墙

双层丝网印刷玻璃 (红色)

玻璃三 (双层中空高遮LOW-E玻璃)

钢化玻璃门

入口立面图

步行街入口

Located in Xinfu District, the project is situated in the heart of Fushun city, with a site area of 138,800m² and gross floor area of 932,100m² including underground floor area of 180,100m² and above-ground floor area of 752,000m², of which, shopping center occupies 105,700m², hotel 32,300m², office 136,400m², residence 255,900m², store 61,500m², urban supporting property 6,500m², moving-back property 160,500m². The project fills in Fushun's blank of large urban complex.

The facade design of podium features Chinese paper-cut for window decoration. It uses hollow-out metal panel and integrates classical architectural culture, trying to echo traditional style by modern materials and methods. Manchu culture and mineral resources are distinctive features of Fushun city, inspired from which, the design

introduces metal panel and elaborately designs modular window paper-cut texture, realizing a perfect integration of classical and modern styles. The hollow-out metal panel forms shadows while the sun shines.

The main entrance design of commerce takes blood amber, which is a rare mineral and what Fushun abounds in, as the theme. It uses screen printing glass to form large-volume door head, attracting customer flow into mall and becoming an eye-catching visual center. The existing indifferent and monotonous surface is changed, while material quality and variety are increased, thus the space experience of the project is greatly improved.

The skylight takes dragon world as the mother theme and the interior design uses dragon scale as the basic element. The whole interior

walking street is just like that a dragon is playing with two beads, while the project is like a lying dragon lurking at the bank of Hunjiang River. The oval atrium is the most important part of the entire space. The side plate is designed like a flying dragon through clouds, partly hidden and partly invisible. The steel structure of the skylight and the rhombus of panoramic elevator also reveal dragon scale element. The round atrium adopts a design method similar to yet different from the oval atrium's. The side plate uses dragon scale pattern and creates texture and sculpture-like shape, looking like a gloriously radiant flying dragon in the light. Instead of using soft and radian lines, the ceiling shape turns to use hard and angular lines, echoing the oval atrium's ceiling. The panoramic elevator features orderly square shape which is evolved from dragon scale.

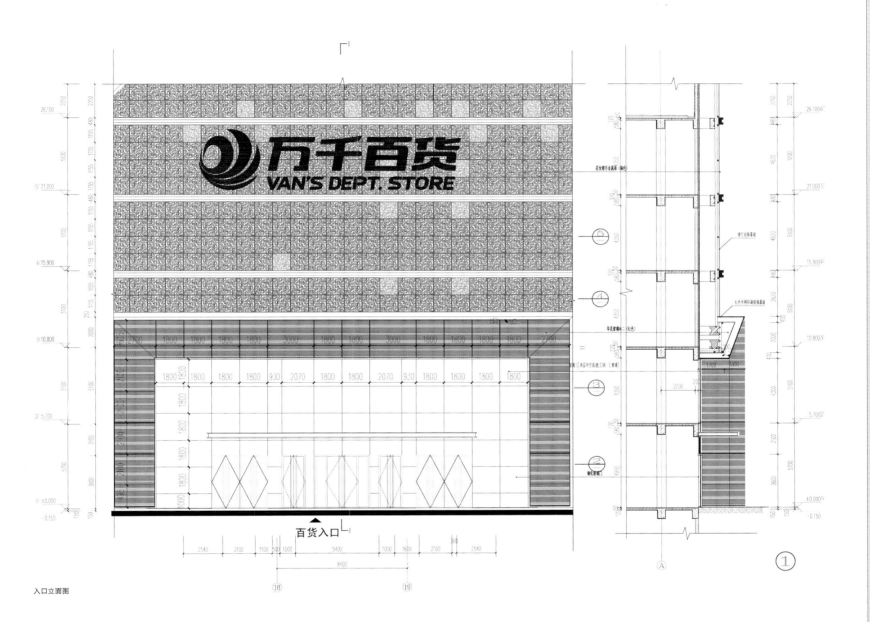

万千百货
VAN'S DEPT. STORE

百货入口

入口立面图

Wanda Realm is grand and dignified, yet concise and bright on facade design. It pays great attention to interacting with urban environment, trying to coordinate building and environment while endow the former with unique characters. The facade employs three-section design method which is commonly used in classical architecture. The top section of the hotel building features dense pilasters, traditional yet fashionable. The middle section uses concise beige stone paint plus deep window openings. The podium section features four-storey high French window, as well as thick stone circling, concise and grand.

The hotel lobby on the first floor boasts a large chandelier with a diameter of 10m on the ceiling, which is themed of rose—the city flower of Fushun, with brilliant crystal outlining florid petals. The main background wall boasts pattern from imperial robe, revealing strong Manchu culture. This not only follows the traditional luxury path of Wanda hotels, but also combines local culture.

抚顺万达广场
Wanda Plaza, Fushun

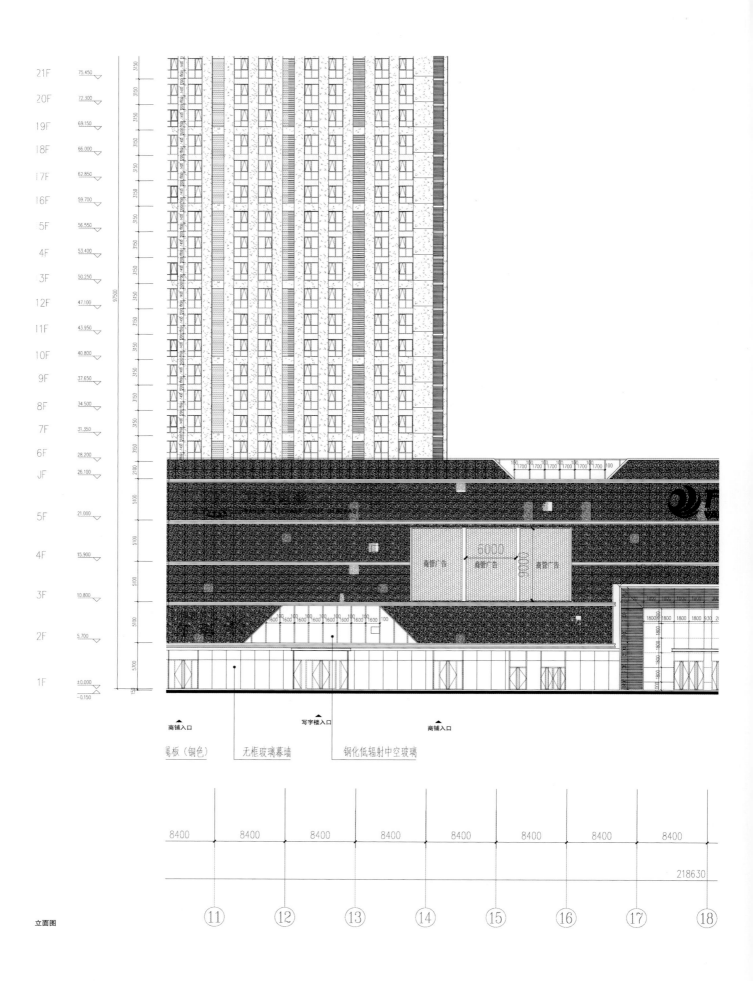

立面图

抚顺万达广场
Wanda Plaza, Fushun

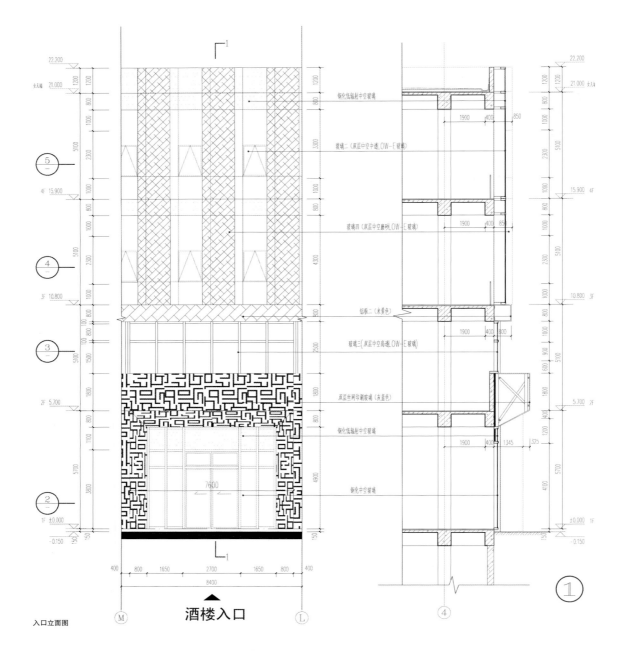

酒楼入口

入口立面图

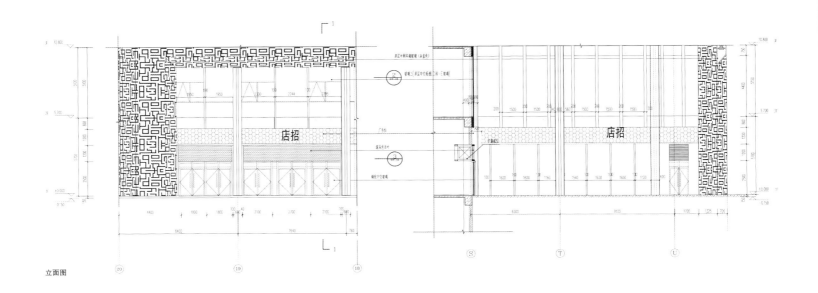

立面图

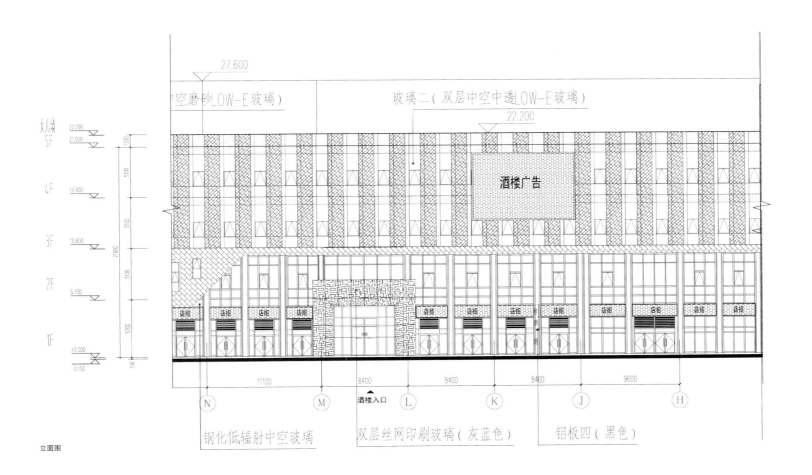

立面图

抚顺万达广场
Wanda Plaza, Fushun

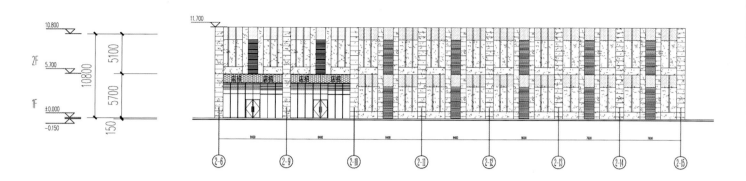

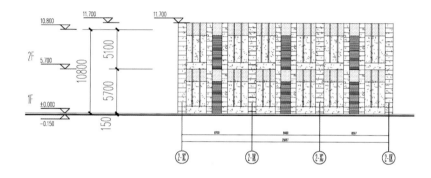

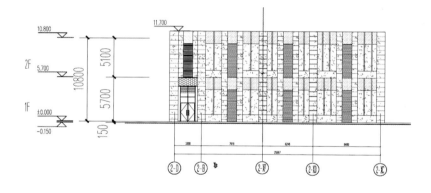

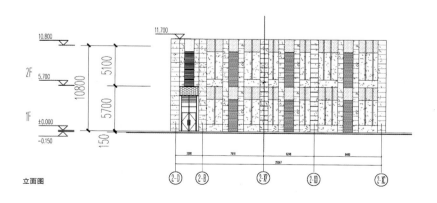

立面图

抚顺万达广场
Wanda Plaza, Fushun

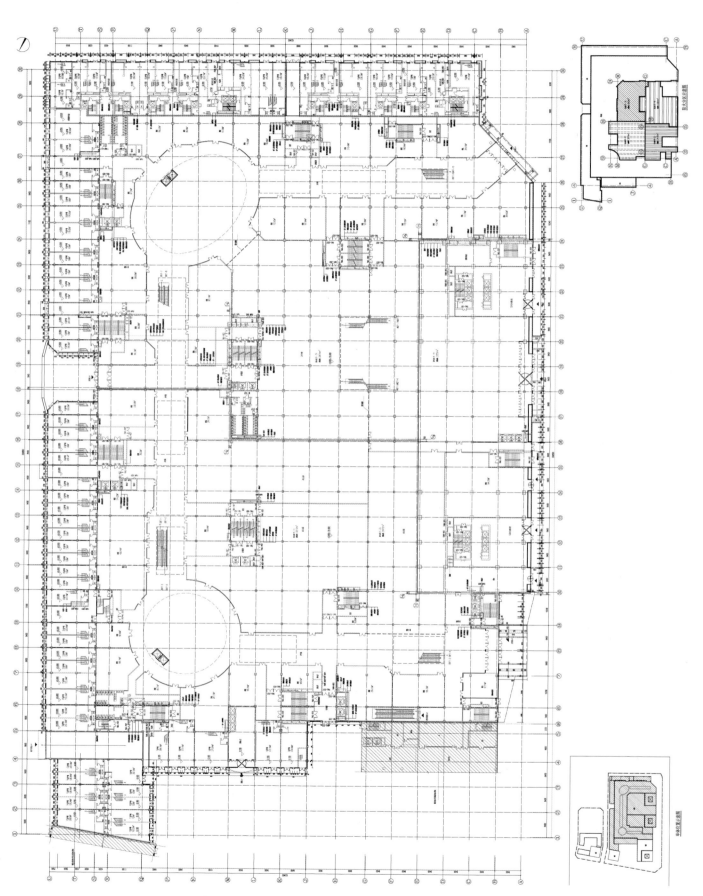

一层平面图

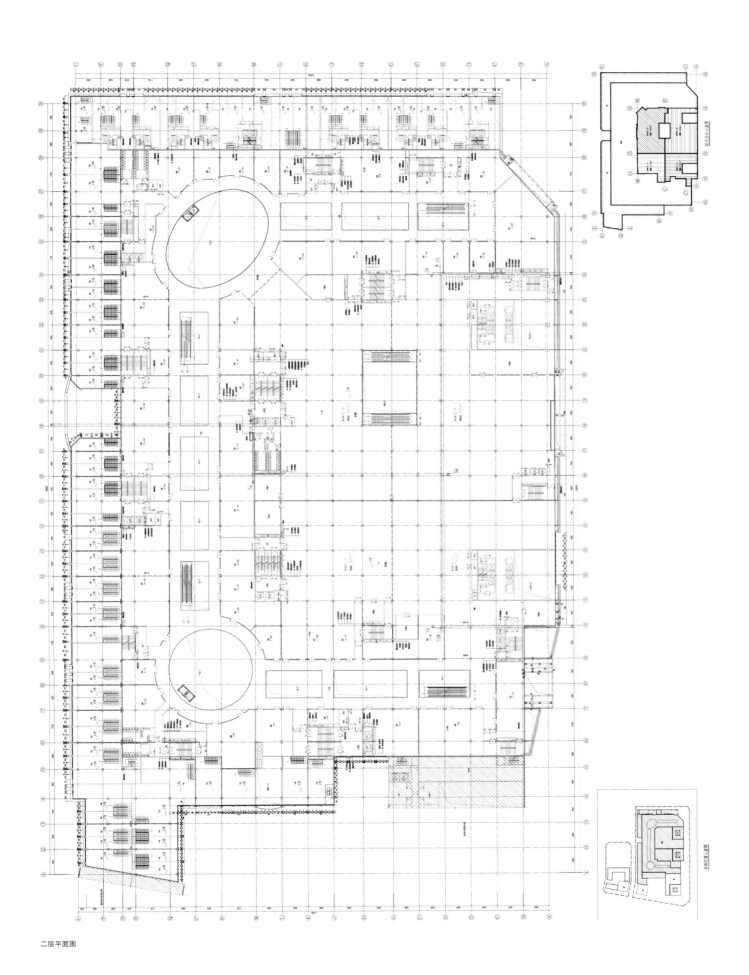

二层平面图

抚顺万达广场
Wanda Plaza, Fushun

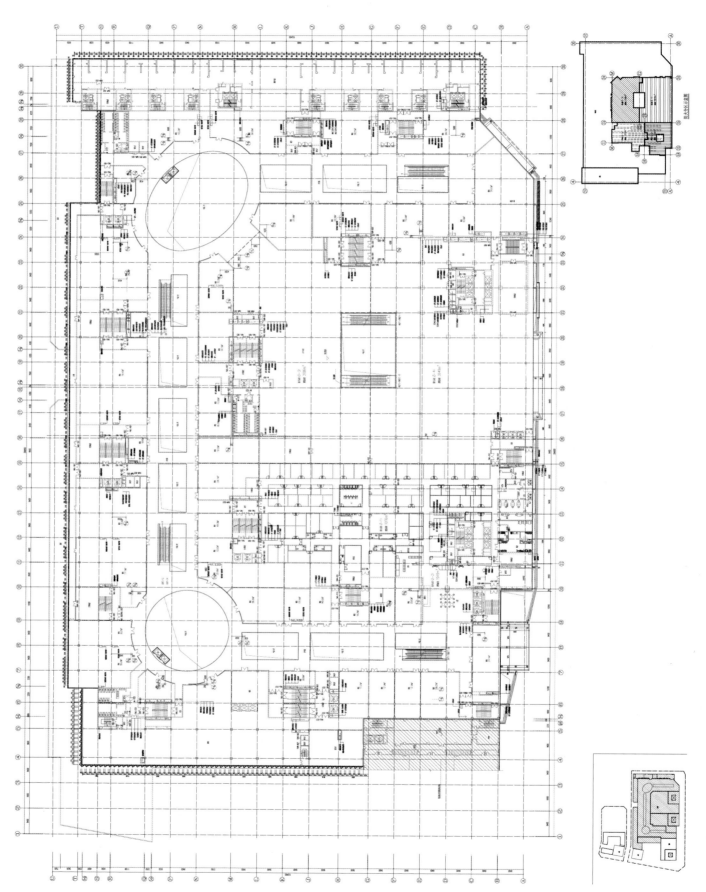

三层平面图

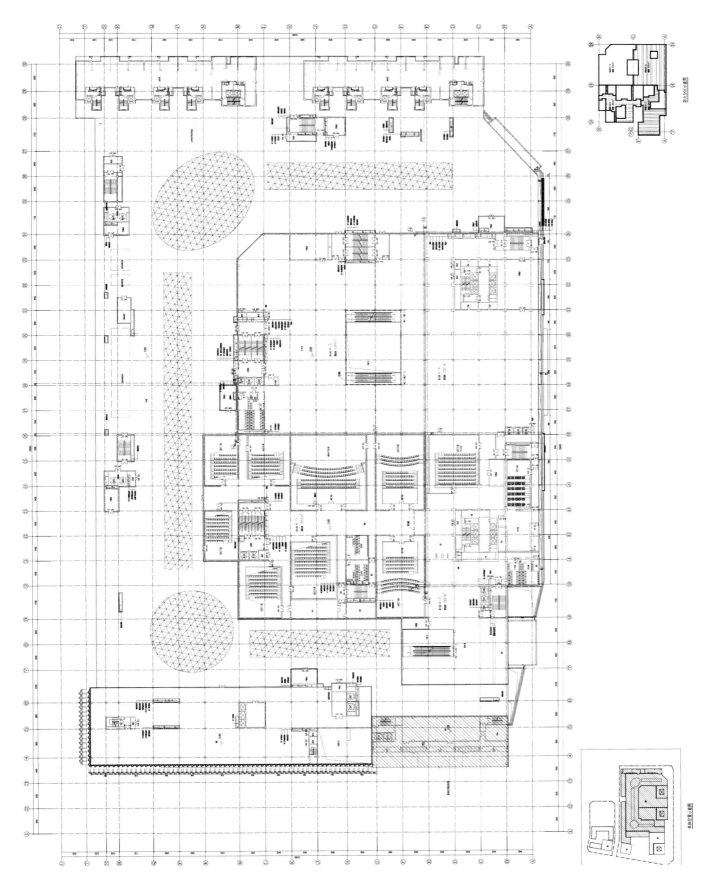

四层平面图

合肥砂之船艺术商业广场
Sasseur Art Commercial Plaza

地点: 安徽省合肥市蜀山区　用地面积: 132 212平方米　建筑面积: 316 106平方米
Location: Shushan District, Hefei, Anhui Province, China　Site Area: 132,212m²　Building Area: 316,106m²

合肥砂之船艺术商业广场
Sasseur Art Commercial Plaza

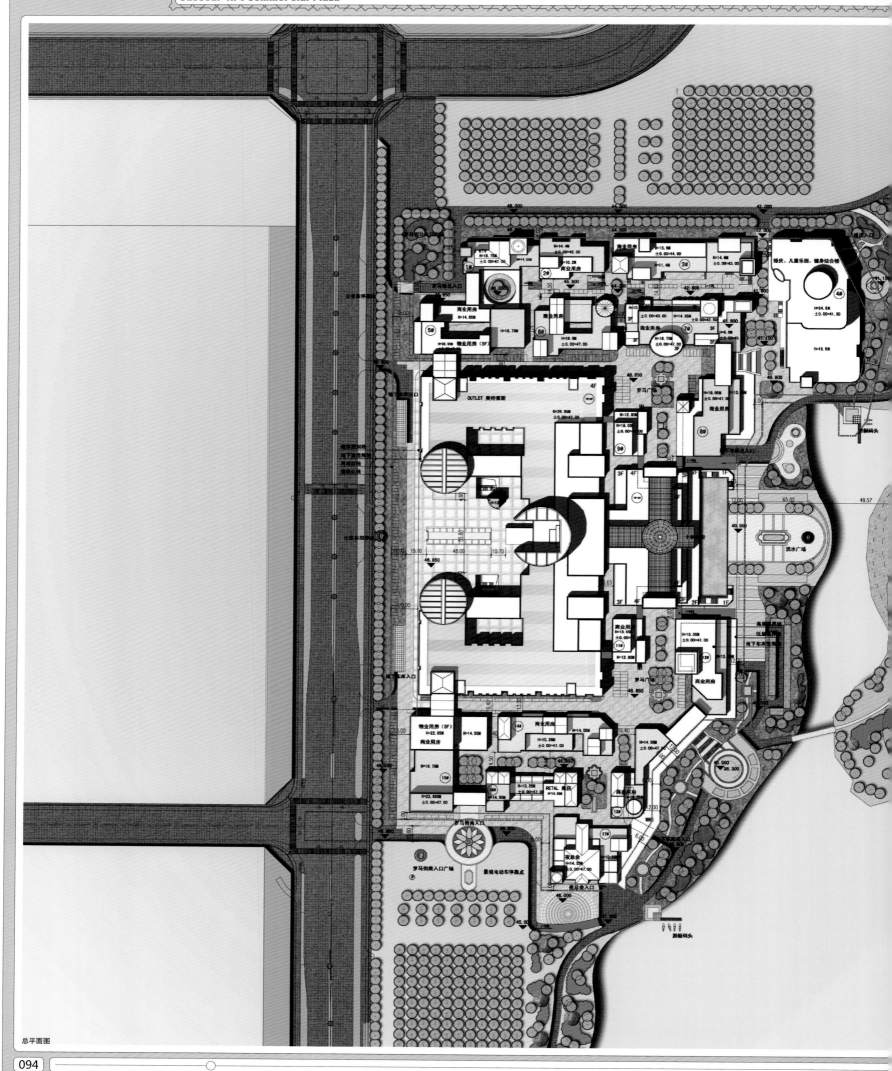

总平面图

地点：安徽省合肥市蜀山区　用地面积：132 212平方米　建筑面积：316 106平方米
Location: Shushan District, Hefei, Anhui Province, China　Site Area: 132,212m²　Building Area: 316,106m²

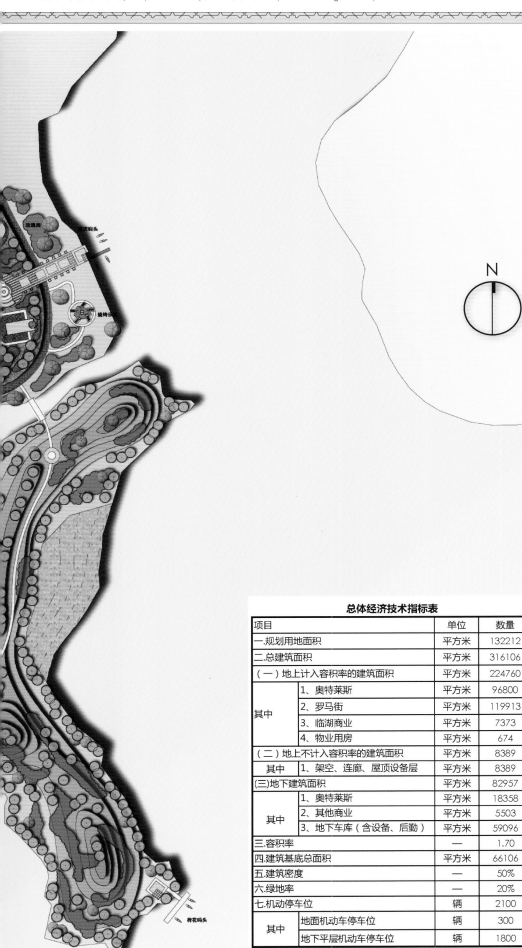

合肥砂之船艺术商业广场位于合肥市蜀山区国家高新技术产业开发区长宁大道以东，彩虹路以南，望江西路以北。用地西侧为长宁大道，并有宽12米的绿化带相隔，东侧紧邻王咀湖景观绿化带。项目规划用地面积为132 212平方米，建筑面积316 106平方米，基地基本呈南北长约500米、东西宽约350米的长方形，地块海拔在38~47米之间，整体呈西高东低，地形高差适中。

本项目的规划理念为"一轴、一心、双环、六节点"："一轴"，由沿长宁大道主广场—奥特莱斯主题建筑—十字金街—滨水广场—王咀湖构成东西方向的主轴线，既是主要的商业动线，也是景观轴线；"一心"，由奥特莱斯主体建筑与十字金街组成的核心建筑群，作为整个项目的重点空间，聚集了国际各大主品牌，是整个项目的中心节点和精神空间；"双环"，环由奥特莱斯主体建筑与周围罗马风情街形成，为基础内部主要机动车流线。另一环由罗马风情街本身形成，为主要购物步行街流线；"六节点"，由奥特莱斯主体建筑前靠近长宁大道一侧广场、南北罗马街入口广场、南北罗马街内广场及东侧临王咀湖滨水广场构成六个景观空间节点构成。

在功能布局上，主体奥特莱斯为地上四层、地下一层的多层建筑。建筑形体按"凹"字形中轴对称布置。地下一层的主要功能为零售及餐饮，并与地下车库相联。地上一至四层设置服装折扣馆和影院等。十字金街为地上四层、地下一层，沿王咀湖逐层退台式的建筑。地下设置3 000平方米左右的精品超市及酒吧餐厅，地上一至二层布置意大利精品服饰馆、品牌珠宝馆等高端主力店，三层至四层为餐饮功能，用餐者可在此尽享王咀湖美景。

总体经济技术指标表

项目		单位	数量
一.规划用地面积		平方米	132212
二.总建筑面积		平方米	316106
（一）地上计入容积率的建筑面积		平方米	224760
其中	1、奥特莱斯	平方米	96800
	2、罗马街	平方米	119913
	3、临湖商业	平方米	7373
	4、物业用房	平方米	674
（二）地上不计入容积率的建筑面积		平方米	8389
其中	1、架空、连廊、屋顶设备层	平方米	8389
（三）地下建筑面积		平方米	82957
其中	1、奥特莱斯	平方米	18358
	2、其他商业	平方米	5503
	3、地下车库（含设备、后勤）	平方米	59096
三.容积率		—	1.70
四.建筑基底总面积		平方米	66106
五.建筑密度		—	50%
六.绿地率		—	20%
七.机动停车位		辆	2100
其中	地面机动车停车位	辆	300
	地下平层机动车停车位	辆	1800

规划总平面图

合肥砂之船艺术商业广场
Sasseur Art Commercial Plaza

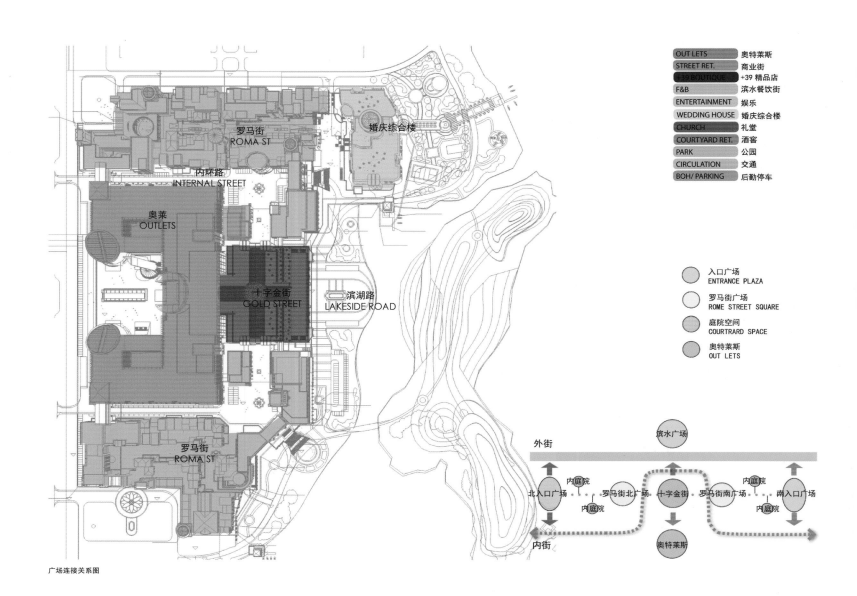

OUT LETS	奥特莱斯
STREET RET.	商业街
+39 BOUTIQUE	+39 精品店
F&B	滨水餐饮街
ENTERTAINMENT	娱乐
WEDDING HOUSE	婚庆综合楼
CHURCH	礼堂
COURTYARD RET.	酒窖
PARK	公园
CIRCULATION	交通
BOH/ PARKING	后勤停车

入口广场 ENTRANCE PLAZA

罗马街广场 ROME STREET SQUARE

庭院空间 COURTRARD SPACE

奥特莱斯 OUT LETS

罗马街 ROMA ST
婚庆综合楼
内环路 INTERNAL STREET
奥莱 OUTLETS
十字金街 GOLD STREET
滨湖路 LAKESIDE ROAD
罗马街 ROMA ST

广场连接关系图

罗马街分为南、北两区,分别布置零售商铺、餐饮及KTV等娱乐业态。罗马街内街采用景观水街的设计,结合两边商铺及三层露台,形成空间丰富的购物休闲空间。婚庆儿童乐园健身综合楼为三至四层建筑:一、二层为婚庆庄园,三层为儿童乐园,四层为健身中心。

在建筑立面上,主体奥特莱斯采用瑞士建筑大师马里奥·博塔的建筑风格,对传统的欧式建筑语言进行精彩的提炼,采用面砖拼贴的立面处理手法,简洁中富含细节,通过尺度、材料的对比,突显建筑质感与品位。而十字金街的建筑风格及空间意向则直接取材于米兰埃马努埃莱二世拱廊,以超尺度的拱廊及穹顶构筑了恢宏的商业空间节点,成为整个项目的核心。罗马街采用了古罗马小镇民居式的建筑风格。建筑多为两至三层民居式风格。屋顶轮廓线高低起伏、错落有致,形成丰富天际线的同时,又与奥莱主体建筑形成微妙的反差和对比。

地点：安徽省合肥市蜀山区　用地面积：132 212平方米　建筑面积：316 106平方米
Location: Shushan District, Hefei, Anhui Province, China　Site Area: 132,212m²　Building Area: 316,106m²

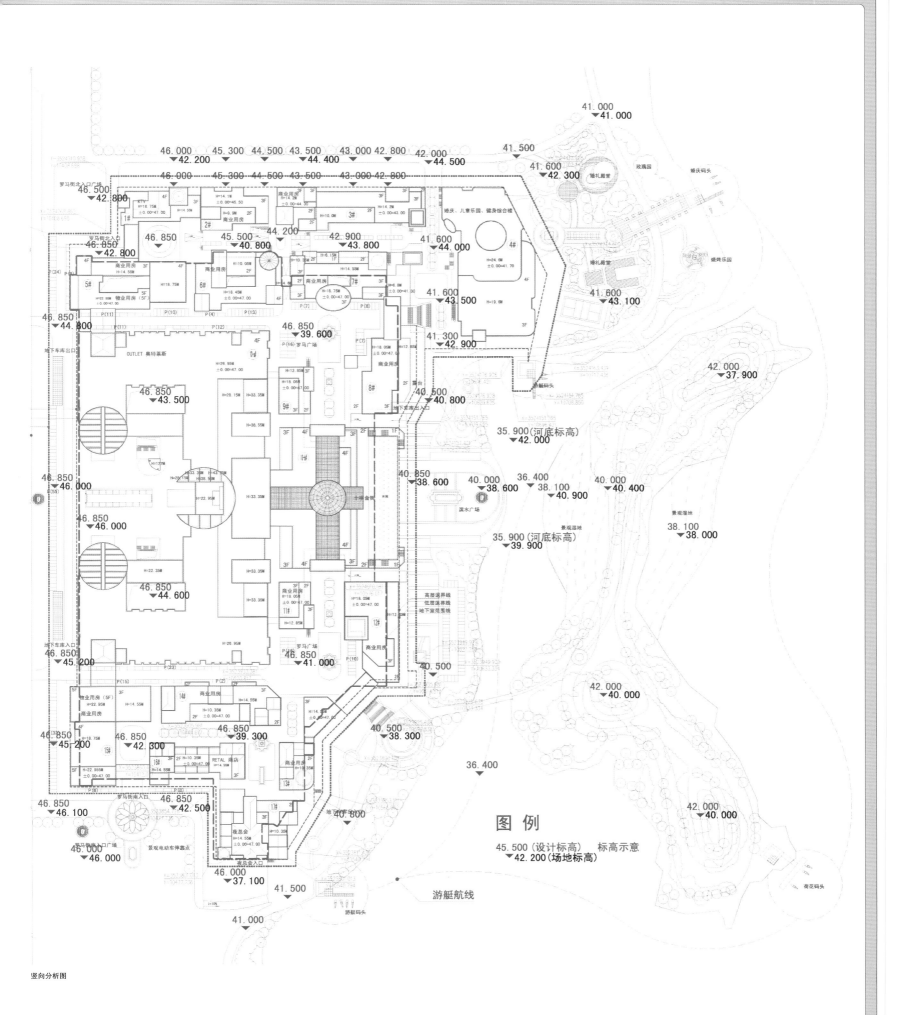

图 例

45.500（设计标高）　标高示意
▼42.200（场地标高）

竖向分析图

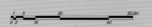

合肥砂之船艺术商业广场
Sasseur Art Commercial Plaza

The project is situated in Hefei High-tech Zone, to the east of Changning Avenue, to the south of Caihong Road, to the north of Wangjiang West Road. The west of the site is Changning Avenue separated by a 12m wide green belt, while the east is close to Wangju Lake green belt. The project has site area of 132,212m² and building area of 316,106m². The site is a rectangle, 500m long from south to north, 350m wide from east to west. It is 38-47m high above sea level, and descends from west to east, with a moderate altitude difference.

The project features a planning concept of "one axis, one heart, two rings, six nodes". One axis: the east-west main axis, formed by Changning Avenue main square, Outlets theme building, cross gold street, waterfront square and Wangju Lake, is not only a main commercial circulation, but also a landscape axis. One heart: the core building cluster, composed of Outlets body building and cross gold street, is the central node and spirit space of the project, gathering numerous international brands. Two rings: one ring is formed by Outlets body building and surrounding Roman-style street, acting as the main motor vehicle circulation inside. The other ring is formed by the Roman-style street, functioning as the main shopping walking street. Six nodes: they refer to the square that is close to Changning Avenue and in front of Outlets body building, the two entrance squares of south and north Roman Street, the two inner squares of south and north Roman Street, the waterfront square facing Wangju Lake.

The Outlets body building has four above-ground floors and one underground floor, designed in the shape of shape. The underground floor mainly houses retail and F&B, and is connected to underground parking. The four above-ground floors are arranged with discount clothes hall and cinema. Cross gold street has four above-ground floors and one underground floor, and it terraces layer by layer along Wangju Lake. The underground floor is equipped with about 3,000m2 boutique supermarket, bar and restaurant. The first two floors are captured by Italian boutique clothes hall, brand jewelry hall and other high-end anchor stores. The top two floors are housed with F&B, enjoying the beautiful view of Wangju Lake.

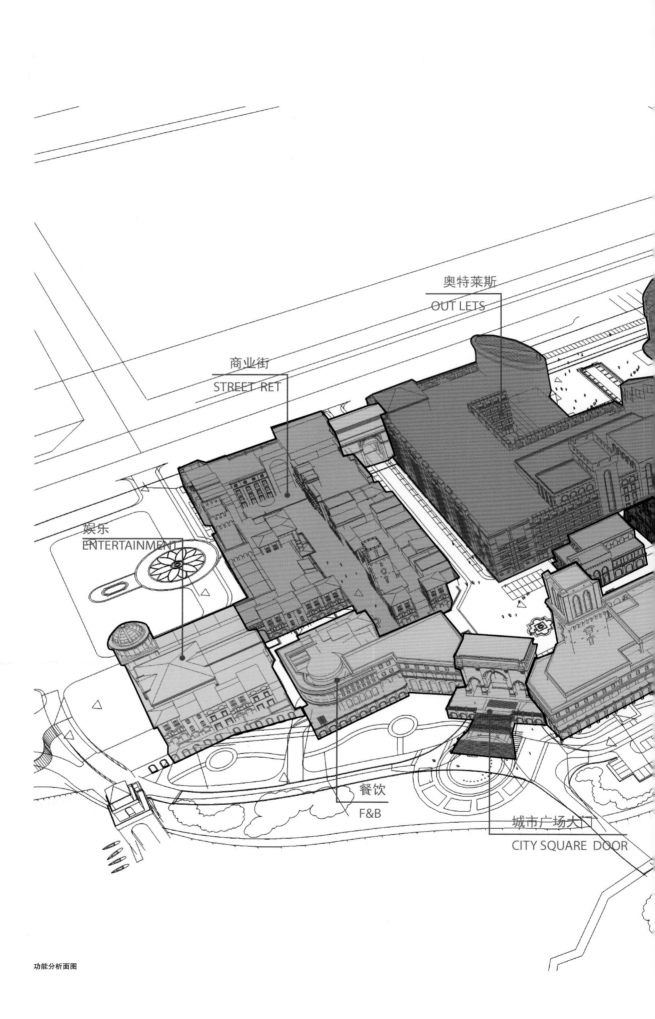

奥特莱斯
OUT LETS

商业街
STREET RET

娱乐
ENTERTAINMENT

餐饮
F&B

城市广场大门
CITY SQUARE DOOR

功能分析面图

地点：安徽省合肥市蜀山区　用地面积：132 212平方米　建筑面积：316 106平方米
Location: Shushan District, Hefei, Anhui Province, China　Site Area: 132,212m²　Building Area: 316,106m²

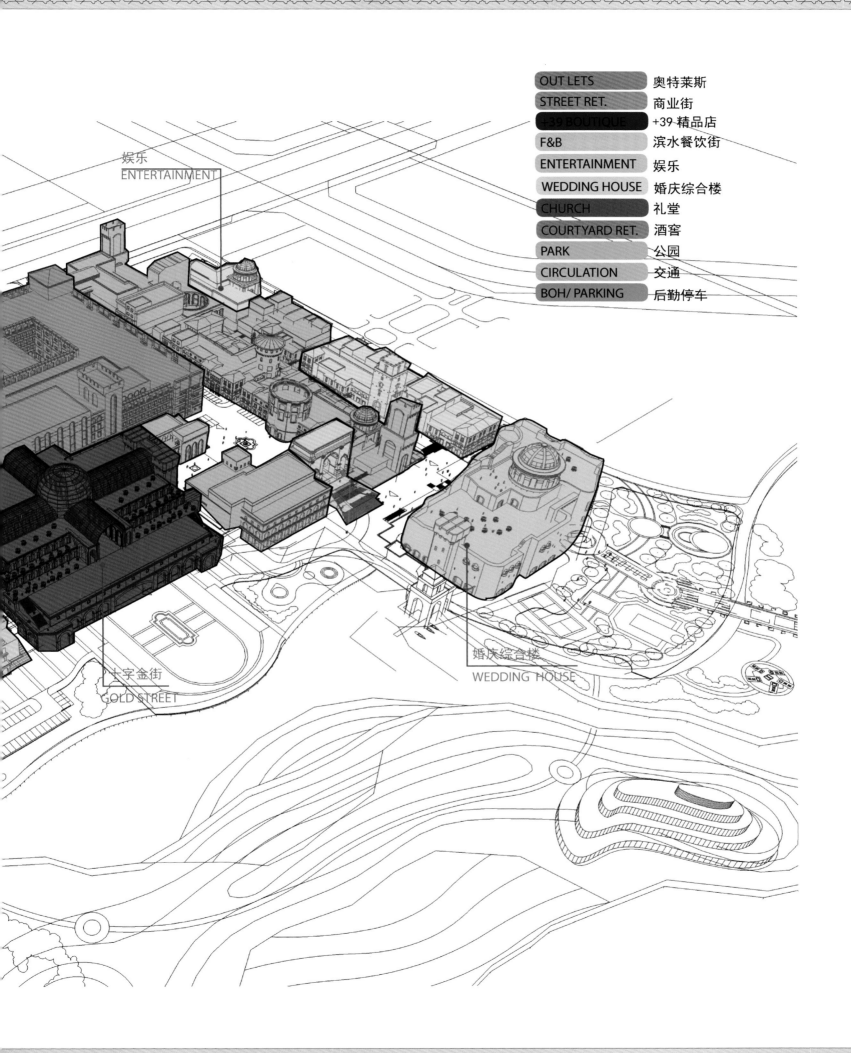

OUT LETS	奥特莱斯
STREET RET.	商业街
+39 BOUTIQUE	+39 精品店
F&B	滨水餐饮街
ENTERTAINMENT	娱乐
WEDDING HOUSE	婚庆综合楼
CHURCH	礼堂
COURTYARD RET.	酒窖
PARK	公园
CIRCULATION	交通
BOH/ PARKING	后勤停车

娱乐
ENTERTAINMENT

十字金街
GOLD STREET

婚庆综合楼
WEDDING HOUSE

合肥砂之船艺术商业广场
Sasseur Art Commercial Plaza

合肥砂之船艺术商业广场
Sasseur Art Commercial Plaza

地点：安徽省合肥市蜀山区　用地面积：132 212平方米　建筑面积：316 106平方米
Location: Shushan District, Hefei, Anhui Province, China　Site Area: 132,212m²　Building Area: 316,106m²

合肥砂之船艺术商业广场
Sasseur Art Commercial Plaza

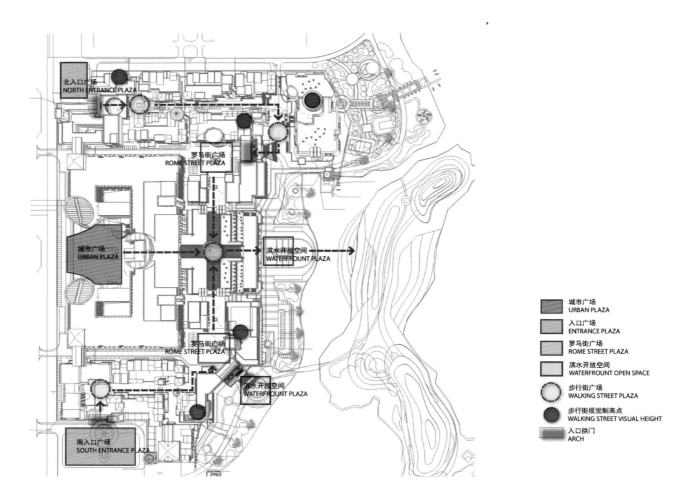

北入口广场 NORTH ENTRANCE PLAZA
罗马街广场 ROME STREET PLAZA
城市广场 URBAN PLAZA
滨水开放空间 WATERFROUNT PLAZA
罗马街广场 ROME STREET PLAZA
滨水开放空间 WATERFROUNT PLAZA
南入口广场 SOUTH ENTRANCE PLAZA

城市广场 URBAN PLAZA
入口广场 ENTRANCE PLAZA
罗马街广场 ROME STREET PLAZA
滨水开放空间 WATERFROUNT OPEN SPACE
步行街广场 WALKING STREET PLAZA
步行街视觉制高点 WALKING STREET VISUAL HEIGHT
入口拱门 ARCH

Roman Street is divided into south and north parts, which are respectively arranged with retail stores and F&B, KTV. Its inner street is of landscaped water street, which is lined with stores and adorned by three-storey terrace, forming a rich shopping and leisure space. The wedding & children's paradise & fitness complex building is four-storey. The first two floors are wedding manor; the third floor is children's paradise; the fourth floor is fitness center.

The facade of Outlets body building features the architectural style of a Swiss architectural master, Mario Botta, smartly extracting essence from traditional European architectural language. It collages facing tile and highlights the quality and taste of the building through the contrast between scale and material. The architectural style and space connotation of cross gold street are directly from Milan Vittorio Emanuele II arcade. The super-scale arcade and dome create a grand commercial space that becomes the core spirit of the project. The Roman Street boasts a folk house style just like in ancient Rome. The buildings are mostly 2 or 3-storey. The contour lines of the roofs enrich the skyline, and meanwhile, contrast with Outlets body building.

地点: 安徽省合肥市蜀山区　用地面积: 132 212平方米　建筑面积: 316 106平方米
Location: Shushan District, Hefei, Anhui Province, China　Site Area: 132,212m²　Building Area: 316,106m²

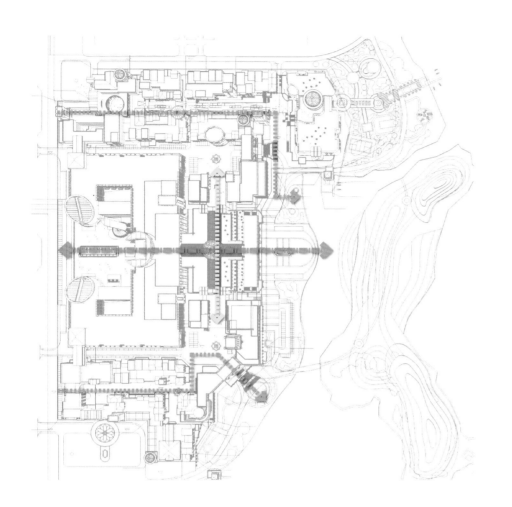

景观主轴
PRIMARY LANDSCAPE AXIS

十字金街景观流线
GOLD STREET RETAIL FLOW

罗马街景观流线
ROME STREET RETAIL FLOW

合肥砂之船艺术商业广场
Sasseur Art Commercial Plaza

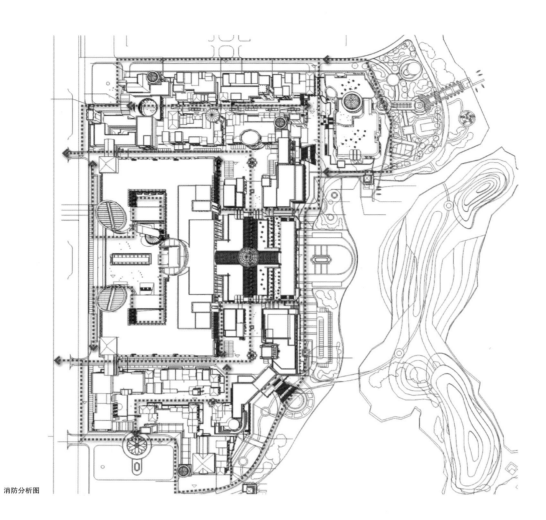

消防分析图

消防环路
FIRE CONTRIL LOOP

地点: 安徽省合肥市蜀山区   用地面积: 132 212平方米   建筑面积: 316 106平方米
Location: Shushan District, Hefei, Anhui Province, China   Site Area: 132,212m²   Building Area: 316,106m²

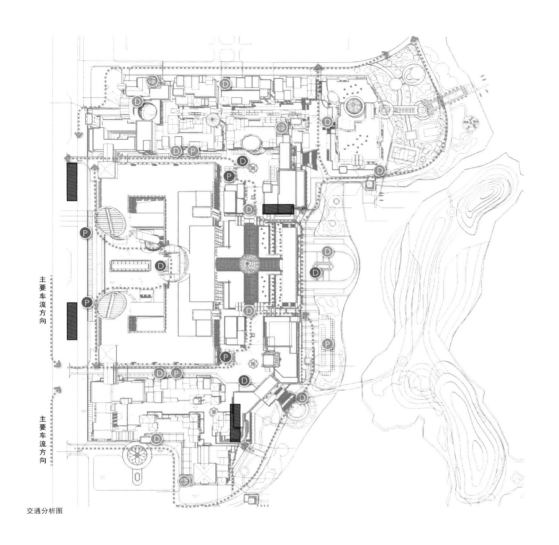

主要车流方向

主要车流方向

地库出入口
UNDERGROUND PARKING

D 奥特莱斯下车点
OUTLETS DROP OFF

D 罗马街下车点
ROMAN STREET DROP OFF

P 奥特莱斯停车
OUTLETS PARKING

P 罗马街停车停车
ROMAN STREET PARKING

车行流线
TRAFFIC CIRCULATION

交通分析图

合肥砂之船艺术商业广场
Sasseur Art Commercial Plaza

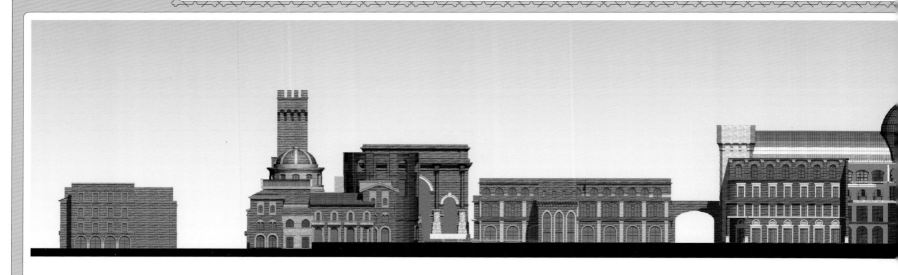

立面图

地点：安徽省合肥市蜀山区　用地面积：132 212平方米　建筑面积：316 106平方米
Location: Shushan District, Hefei, Anhui Province, China　Site Area: 132,212m²　Building Area: 316,106m²

婚庆楼立面图

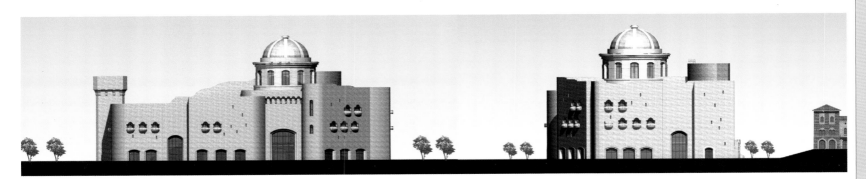

婚庆楼立面图

合肥砂之船艺术商业广场
Sasseur Art Commercial Plaza

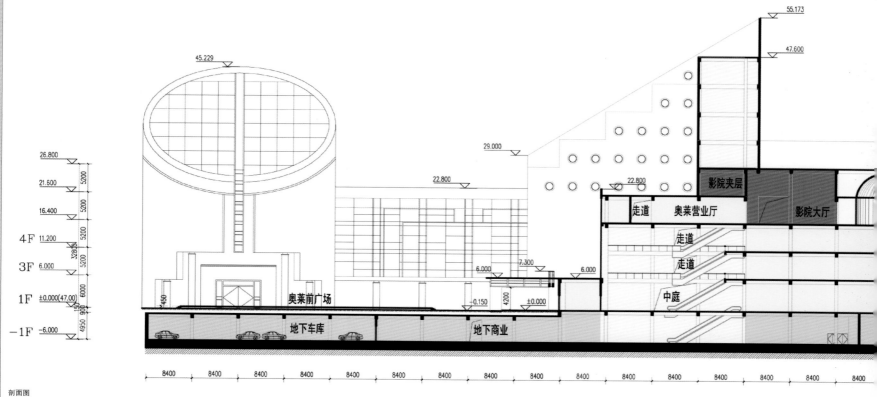

剖面图

地点：安徽省合肥市蜀山区　用地面积：132 212平方米　建筑面积：316 106平方米
Location: Shushan District, Hefei, Anhui Province, China　Site Area: 132,212m²　Building Area: 316,106m²

剖面图

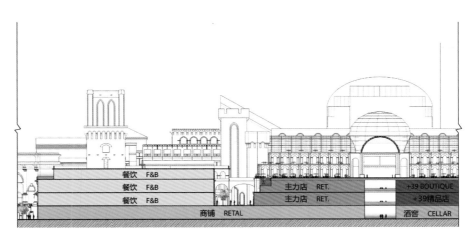

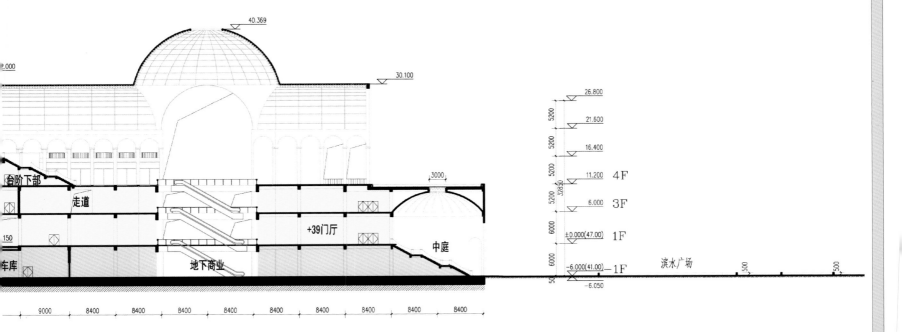

合肥砂之船艺术商业广场
Sasseur Art Commercial Plaza

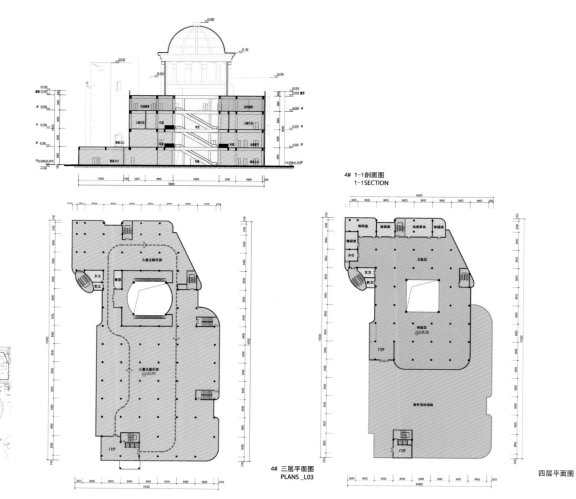

4# 1-1剖面图
1-1SECTION

4# 三层平面图
PLANS_L03

四层平面图

地点：安徽省合肥市蜀山区　用地面积：132 212平方米　建筑面积：316 106平方米
Location: Shushan District, Hefei, Anhui Province, China　Site Area: 132,212m²　Building Area: 316,106m²

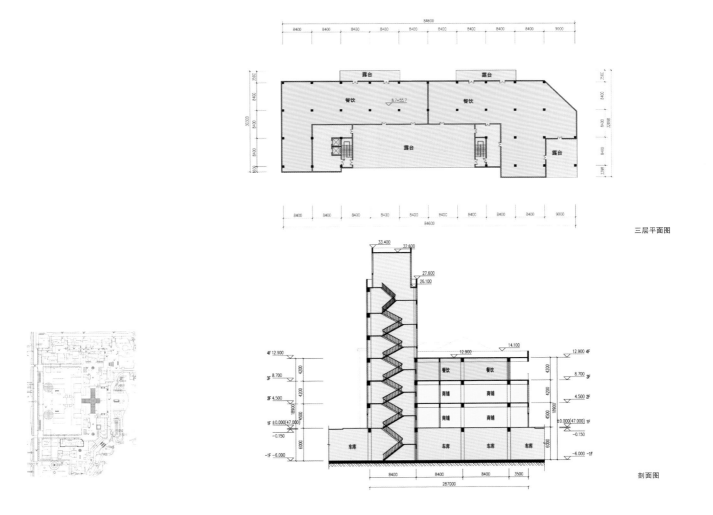

三层平面图

剖面图

合肥砂之船艺术商业广场
Sasseur Art Commercial Plaza

滨水广场透视图

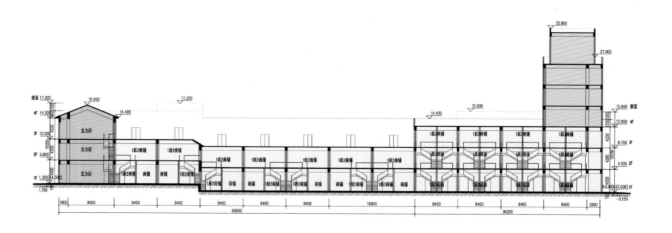

剖面图

地点：安徽省合肥市蜀山区　用地面积：132 212平方米　建筑面积：316 106平方米
Location: Shushan District, Hefei, Anhui Province, China　Site Area: 132,212m²　Building Area: 316,106m²

南罗马广场透视图

合肥砂之船艺术商业广场
Sasseur Art Commercial Plaza

十字金街透视图

合肥砂之船艺术商业广场

十字金街透视图

地点：安徽省合肥市蜀山区　用地面积：132 212平方米　建筑面积：316 106平方米

Location: Shushan District, Hefei, Anhui Province, China　Site Area: 132,212m²　Building Area: 316,106m²

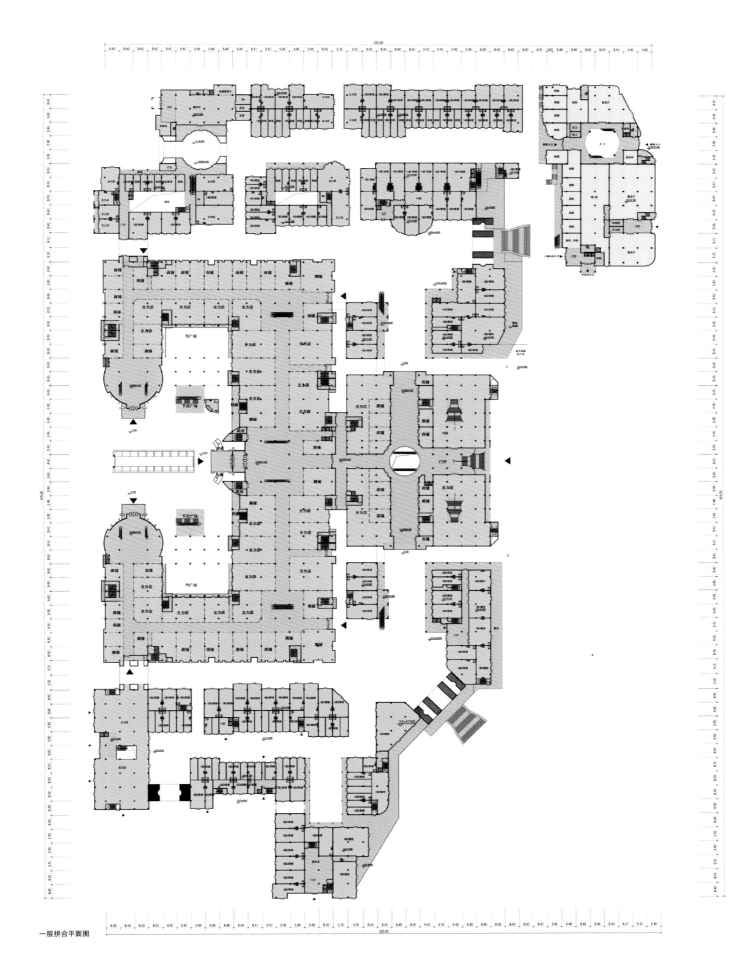

一层拼合平面图

合肥砂之船艺术商业广场
Sasseur Art Commercial Plaza

二层拼合平面图

地点：安徽省合肥市蜀山区　用地面积：132 212平方米　建筑面积：316 106平方米
Location: Shushan District, Hefei, Anhui Province, China　Site Area: 132,212m²　Building Area: 316,106m²

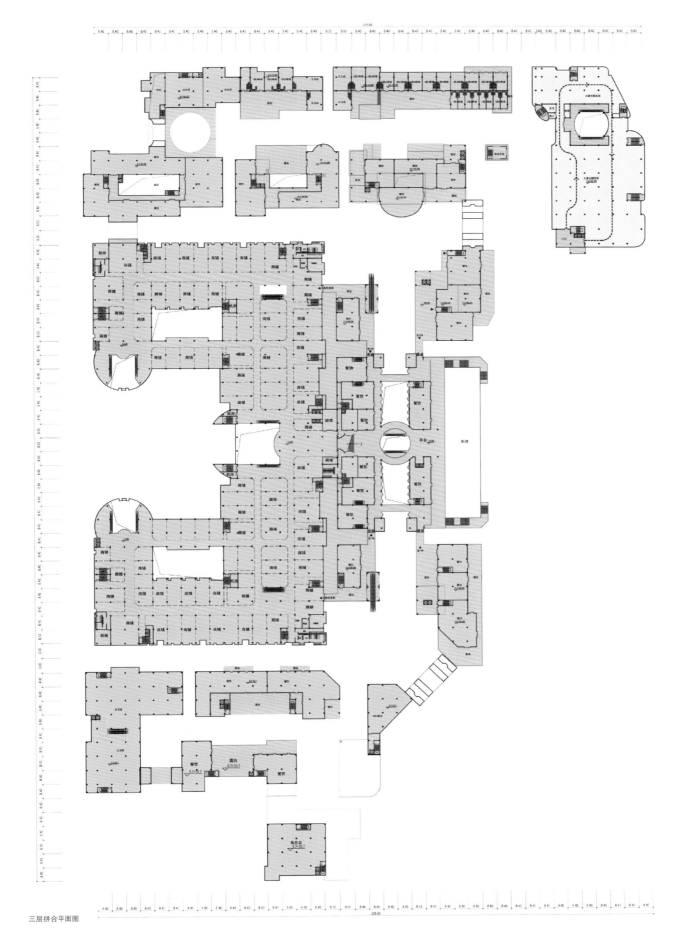

三层拼合平面图

合肥砂之船艺术商业广场
Sasseur Art Commercial Plaza

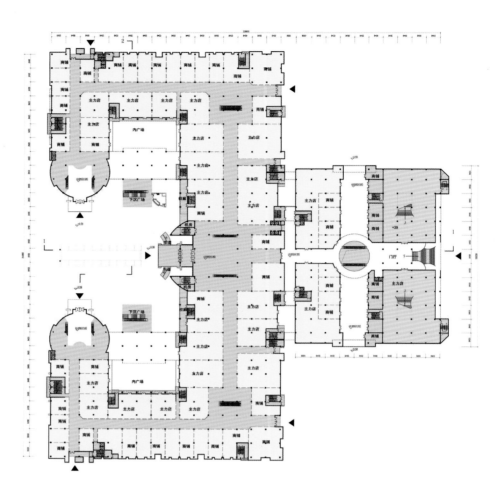

奥莱一层平面图

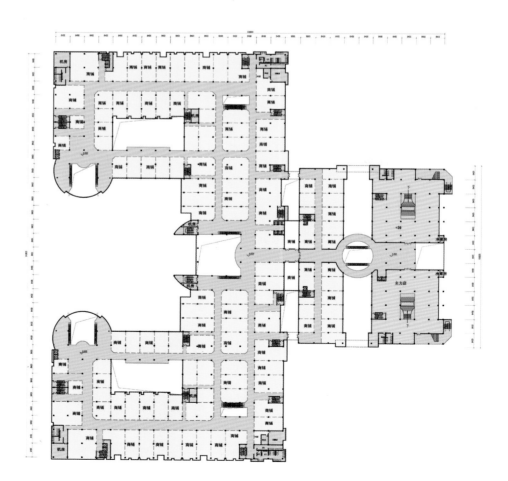

奥莱二层平面图

地点：安徽省合肥市蜀山区　用地面积：132 212平方米　建筑面积：316 106平方米
Location: Shushan District, Hefei, Anhui Province, China　Site Area: 132,212m²　Building Area: 316,106m²

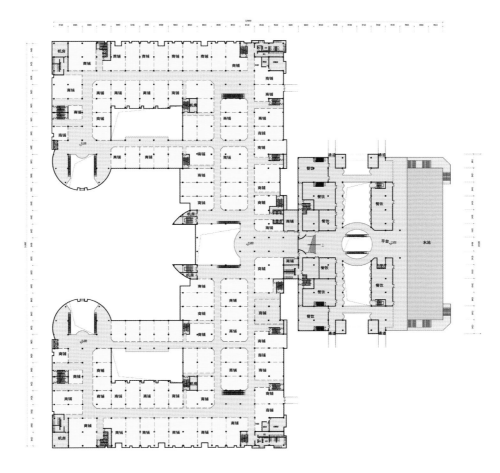

奥莱三层平面图

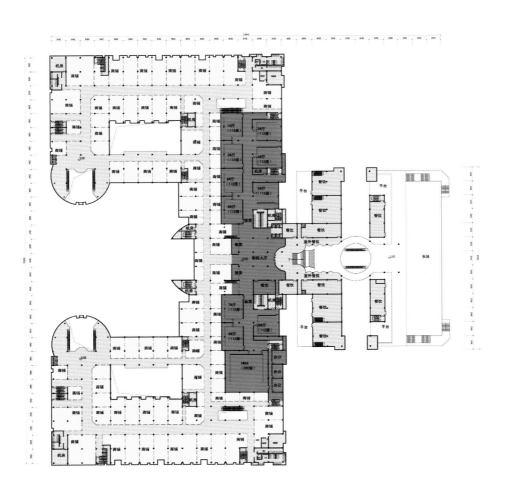

奥莱四层平面图

圣塔莫尼卡广场
Santa Monica Place

圣塔莫尼卡广场
Santa Monica Place

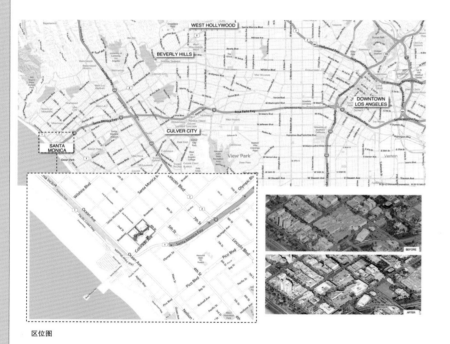

区位图

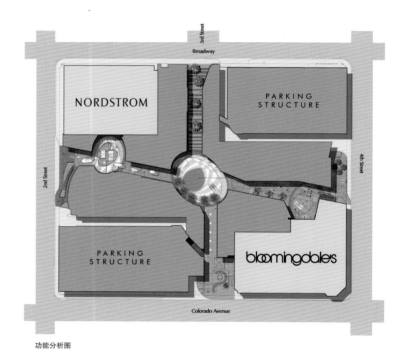

功能分析图

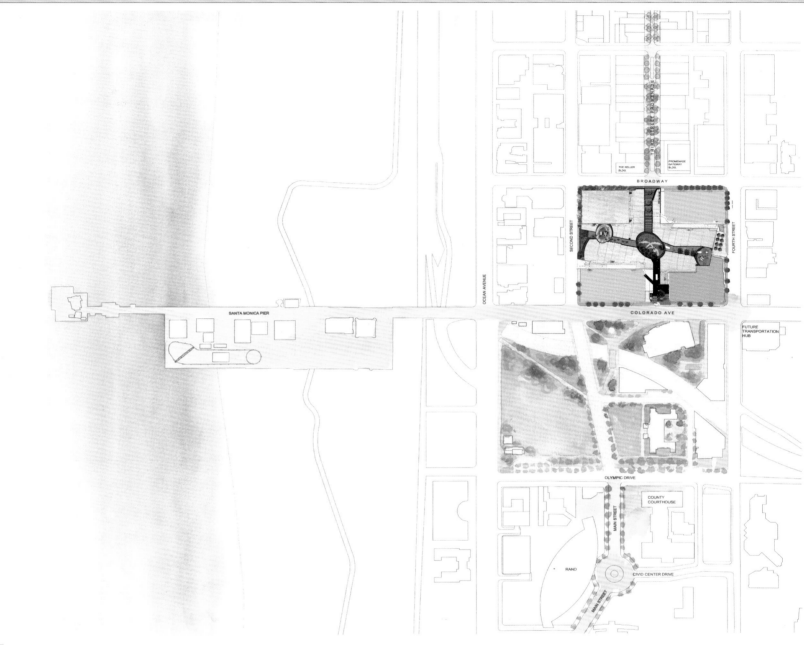

总平面图

新建的圣塔莫尼广场将一个老的封闭的多级商城改造成一个生机勃勃的城市中心，文化上、地理上的开放使得圣塔莫尼广场成为一个适合居住、工作、游玩和参观的极好的地方。

设计师着手建立一个亲切而有活力的公共场所，而不仅仅是一个购物中心。通过将传统商场的城市建设原则与零售设计有机结合，设计师仔细而精心地将这个项目变成今日的城市建筑。

事务所将购物商场的屋顶打开，从而营造宽敞的城市空间，并建立延伸至第三大道步行街的通道。无论从感官欣赏方面还是从其实用功能方面，设计师新加强了项目的自然的感觉，使其能够与圣塔莫尼卡的海边乡村风相融合。

新建的圣塔莫尼卡取代了原来毫无生机的购物商场，充满活力的露天圣塔莫尼卡广场成

为繁荣的圣塔莫尼卡市中心中极其重要的空间。通过天衣无缝的连接增强它的流行度和多用途性，营造新的公共空间，吸引众多的零售租户，包括新的百货公司，它将成为圣塔莫尼卡的中心。

为与城市的户外定位和亲切尺度相协调，捷得事务所旨在创造一个焕然一新的购物中心。在撤掉屋顶并暴露其室内之后，设计师计划让项目延续第三大道步行街的活力，连接现有步行街通道，创建公共广场。从而，项目统一了城市的核心空间，巩固了城市的"心脏"位置，并为居民、上班族、游客创造了新的公共空间。

位于项目四面的四条步行街营造了一种城市的氛围，并捕获了光线和海洋微风。通过将行人引至中央广场，四条步行街通道有利于项目成为一个城市的中央枢纽。每一个入口都体现了

其相邻环境的特征。主要的北入口从视觉上和功能上延续了第三大道步行街的活力。第二大道步行街上的西入口采用有机形体，色彩柔和，呈开放式布局。离行政大楼和市政中心最近的南入口在其特征上则较为正式；东入口是一条连接第四大道步行街餐饮和商务及以东方向住宅的城市廊道。

步行街通道在中心聚集，为公共广场带来了一种新的公共体验。该公共广场是项目的一个标志元素，作为项目的中心，提供了个后现代的聚集之地，可举行表演、艺术展览和其他当地活动。广场椭圆的外形与上部零售层的偏移朝向相结合，使项目所有楼层尽可能多地接收日光。

为强化项目与城市天然相融之感，设计师仔细研究了圣塔莫尼卡市的现有尺度和本土材料，并将它们融入设计中。项目运用赤陶土、玻璃、

石材、砖和不锈钢等材料，通过独特的花纹、不同的空间开口和采光，使得街景生机勃勃。项目所用材料反映了加利福尼亚户外的生活方式。颜色各异、肌理不同的灰红砖、石灰石、瓷砖、木材等让各个空间温暖起来。玻璃百叶窗，外围硬木窗框强化了室内和户外氛围。第三大道步行街上的棕榈树和黄檀元素也被用于广场的设计之中。

圣塔莫尼卡广场
Santa Monica Place

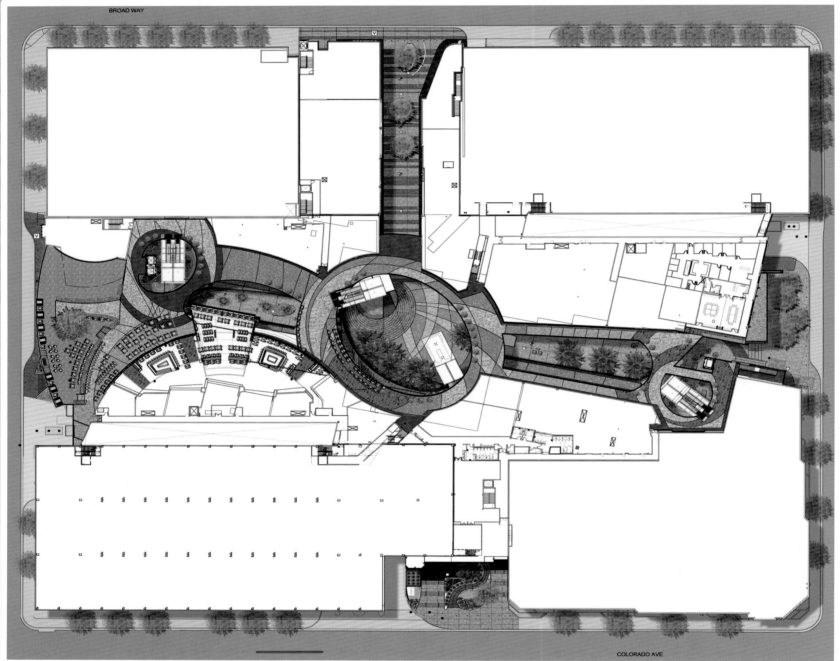

景观分析图

圣塔莫尼卡广场
Santa Monica Place

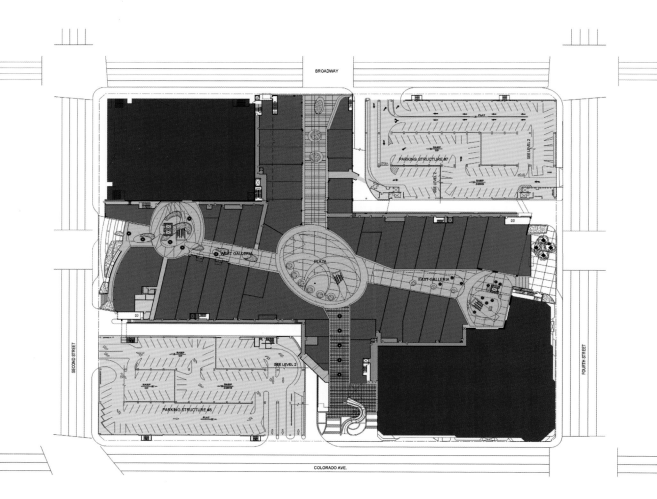

平面布置图

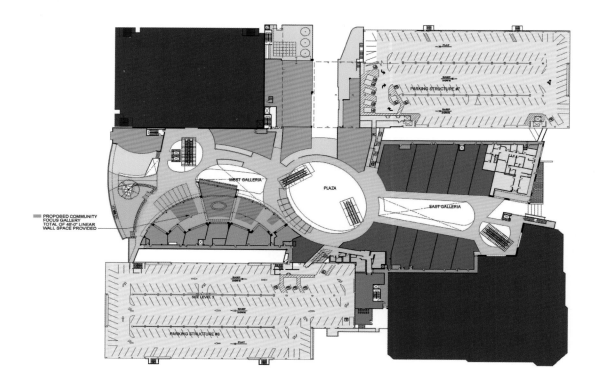

平面布置图

圣塔莫尼卡广场
Santa Monica Place

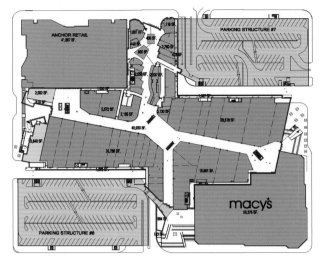
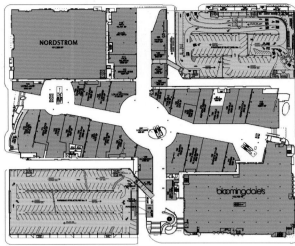

平面布置图

BEFORE

AFTER

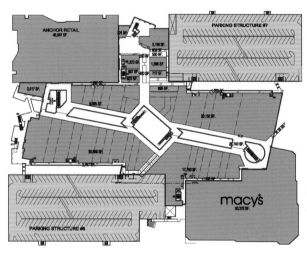
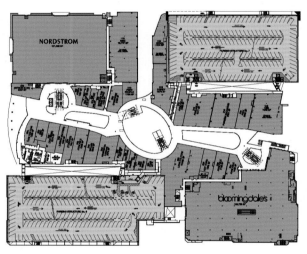

平面布置图

BEFORE

AFTER

LEGEND

RETAIL ANCHOR
RETAIL
F&B
RESTAURANT DEDICATED SEATING
COMMUNITY ROOM
PARKING

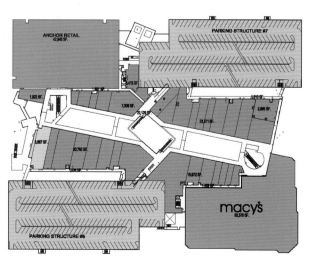
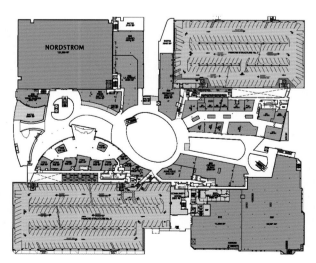

平面布置图

BEFORE

AFTER

圣塔莫尼卡广场
Santa Monica Place

圣塔莫尼卡广场
Santa Monica Place

圣塔莫尼卡广场
Santa Monica Place

BEFORE

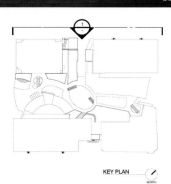

KEY PLAN
NORTH

AFTER

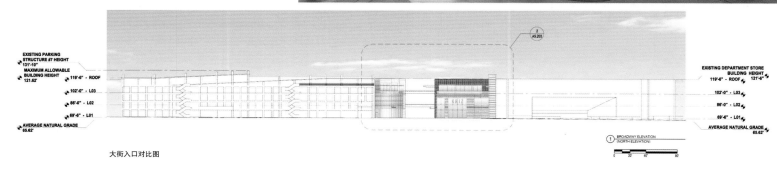

EXISTING PARKING
STRUCTURE #7 HEIGHT
131'-10"
MAXIMUM ALLOWABLE
BUILDING HEIGHT
121.62'
119'-6" - ROOF
102'-0" - L03
86'-0" - L02
69'-6" - L01
AVERAGE NATURAL GRADE
65.62'

EXISTING DEPARTMENT STORE
BUILDING HEIGHT
121'-6"
119'-6" - ROOF
102'-0" - L03
86'-0" - L02
69'-6" - L01
AVERAGE NATURAL GRADE
65.62'

BROADWAY ELEVATION
(NORTH ELEVATION)

大街入口对比图

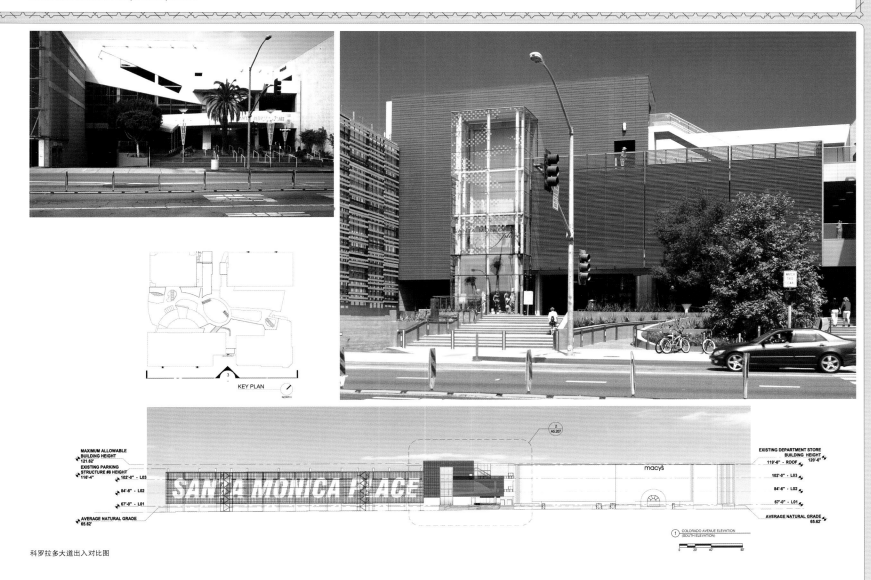

科罗拉多大道出入对比图

1 5   20   50.0m
0 2 10   40

圣塔莫尼卡广场
Santa Monica Place

BEFORE

THE MARKET
True Foo
AFTER

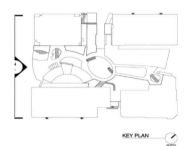

KEY PLAN   NORTH

MAXIMUM ALLOWABLE
BUILDING HEIGHT
121.62'
EXISTING DEPARTMENT STORE
BUILDING HEIGHT
121'-6"
119'-6" - ROOF
102'-0" - L03
84'-6" - L02
67'-0" - L01
AVERAGE NATURAL GRADE
65.62'

MAXIMUM SKYLIGHT HEIGHT
121'-4"
EXISTING PARKING
STRUCTURE #6 HEIGHT
116'-4"
102'-0" - L03
84'-6" - L02
67'-0" - L01
AVERAGE NATURAL GRADE
65.62'

2ND STREET ELEVATION
(WEST ELEVATION)
0   20'   40'   80'

第二街入口对比图

macy's
BEFORE

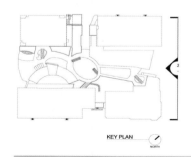

KEY PLAN   NORTH

CB2
CB2
AFTER

MAXIMUM ALLOWABLE
BUILDING HEIGHT - 121.62'
EXISTING DEPARTMENT STORE
BUILDING HEIGHT - 121'-6"
119'-6" - ROOF
102'-0" - L03
84'-6" - L02
67'-0" - L01
AVERAGE NATURAL GRADE
65.62'

macy's

MAXIMUM ALLOWABLE
BUILDING HEIGHT - 121.62'
119'-6" - ROOF
102'-0" - L03
84'-6" - L02
67'-0" - L01
AVERAGE NATURAL GRADE
65.62'

4TH STREET ELEVATION
(EAST ELEVATION)

第四街入口对比图

圣塔莫尼卡广场
Santa Monica Place

阿布扎比阿尔达尔中央市场
Aldar Central Market, Abu Dhabi

地点：**阿拉伯联合酋长国阿布扎比** 占地面积：**57 100平方米** 业主：**ALDAR Properties PJSC** 建筑设计：**福斯特建筑事务所**
Location: Abu Dhabi, United Arab Emirates Site Area: 57,100m² Client: ALDAR Properties PJSC Architecture Design: Foster + Partners

阿布扎比阿尔达尔中央市场
Aldar Central Market, Abu Dhabi

中央市场是阿布扎比最古老的区域之一。该项目的设计灵感源自传统建筑，设计旨在彻底改造这个市场空间，为阿布扎比创造一个新的市民中心。通过对该国际化的购物中心创造一个替代品，设计对项目进行了一次独具特色的现代诠释。作为一种购物体验，设计将奢侈品、精品店和美食市场、工艺品贸易相结合。像传统露天市场一样，这些不同的体验被集合于一个室内建筑中，建筑内充满明媚的阳光、明亮的色彩、美丽的喷泉，广场、庭院和小巷十分和谐。

阿布扎比的气候一年有六个月都是非常舒适宜人的，可以随意漫步或闲坐于室外。设计师受此启发，设计了一系列公共小径和公共广场，移除了室内和室外之间的阻隔。这些新的空间为城市在节庆时期提供了一个重要的中央场地。在一年中的其他六个月，这些空间可通过可滑动的屋面板闭合，达到调节室内温度的目的。屋面板和室内面板上的孔眼在室外得以延续，形成有肌理的立面。面板的设计采用了八角形，同时参考了传统的铺砖和几何学。延续阿布扎比的绿化，项目绿化面积非常之大，裙楼的屋顶设计了一系列梯台式花园。裙楼之上便是高楼群，高度、体量和功能不一，包括办公、公寓、酒店和服务式公寓结合体等。视觉上，建筑非常和谐而统一，立面光滑且反光，即使在这个尘土飞扬的沙漠环境中，后期保养也十分方便。高楼的内部遮阳层可控制眩光和日光。

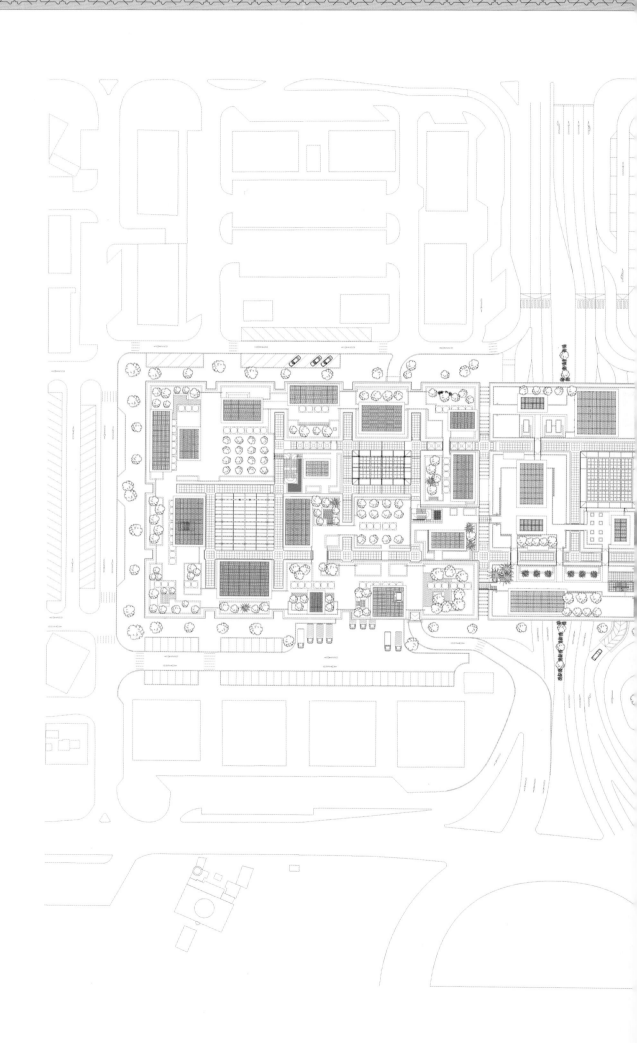

地点：阿拉伯联合酋长国阿布扎比　占地面积：57 100平方米　业主：ALDAR Properties PJSC　建筑设计：福斯特建筑事务所
Location: Abu Dhabi, United Arab Emirates　Site Area: 57,100m²　Client: ALDAR Properties PJSC　Architecture Design: Foster + Partners

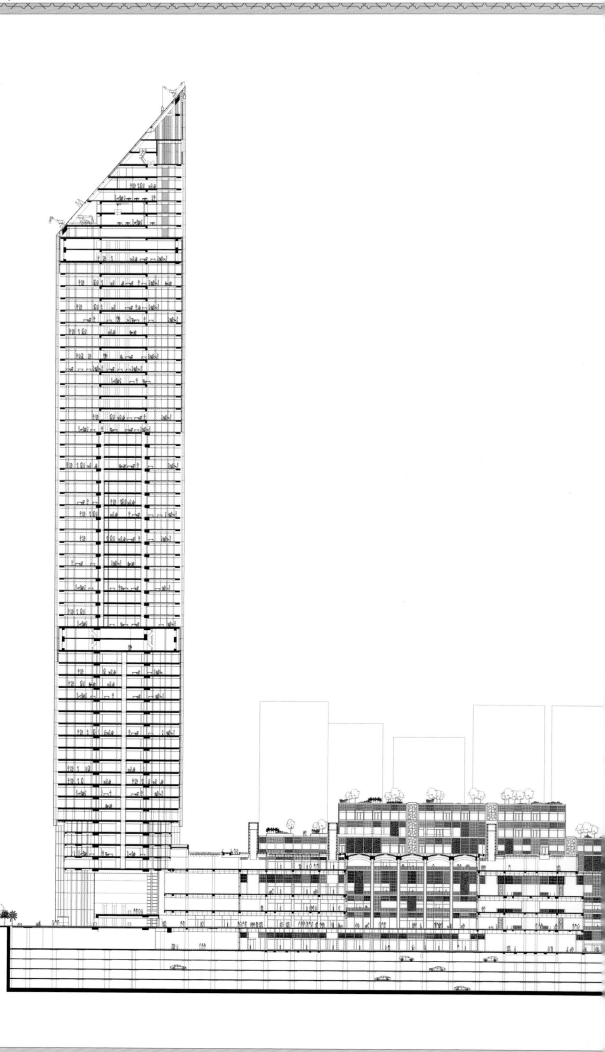

立面图

地点：阿拉伯联合酋长国阿布扎比　占地面积：57 100平方米　业主：ALDAR Properties PJSC　建筑设计：福斯特建筑事务所
Location: Abu Dhabi, United Arab Emirates　Site Area: 57,100m²　Client: ALDAR Properties PJSC　Architecture Design: Foster + Partners

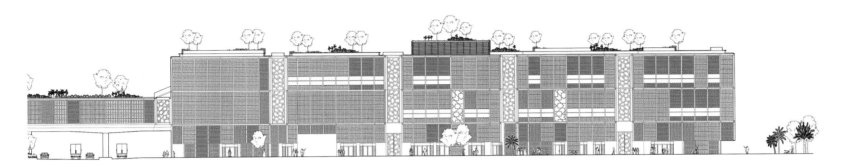

立面图

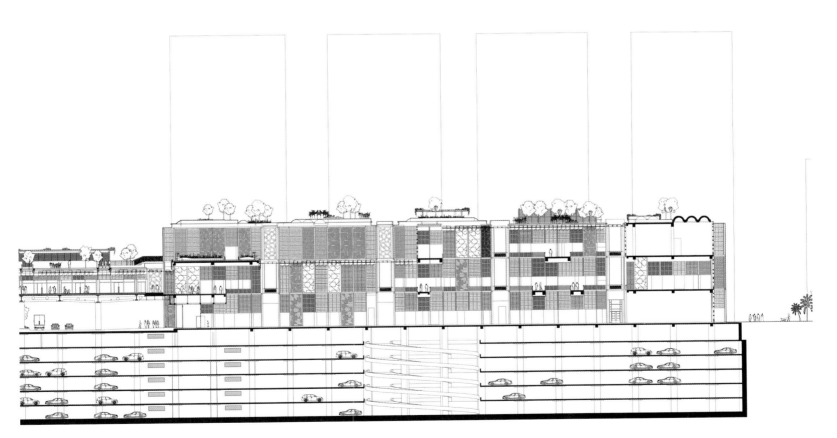

剖面图

阿布扎比阿尔达尔中央市场
Aldar Central Market, Abu Dhabi

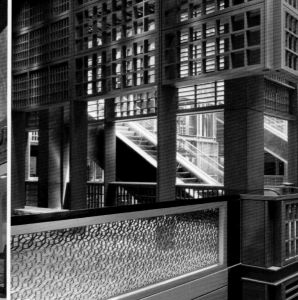

地点：阿拉伯联合酋长国阿布扎比　占地面积：57 100平方米　业主：ALDAR Properties PJSC　建筑设计：福斯特建筑事务所
Location: Abu Dhabi, United Arab Emirates　Site Area: 57,100m²　Client: ALDAR Properties PJSC　Architecture Design: Foster + Partners

阿布扎比阿尔达尔中央市场
Aldar Central Market, Abu Dhabi

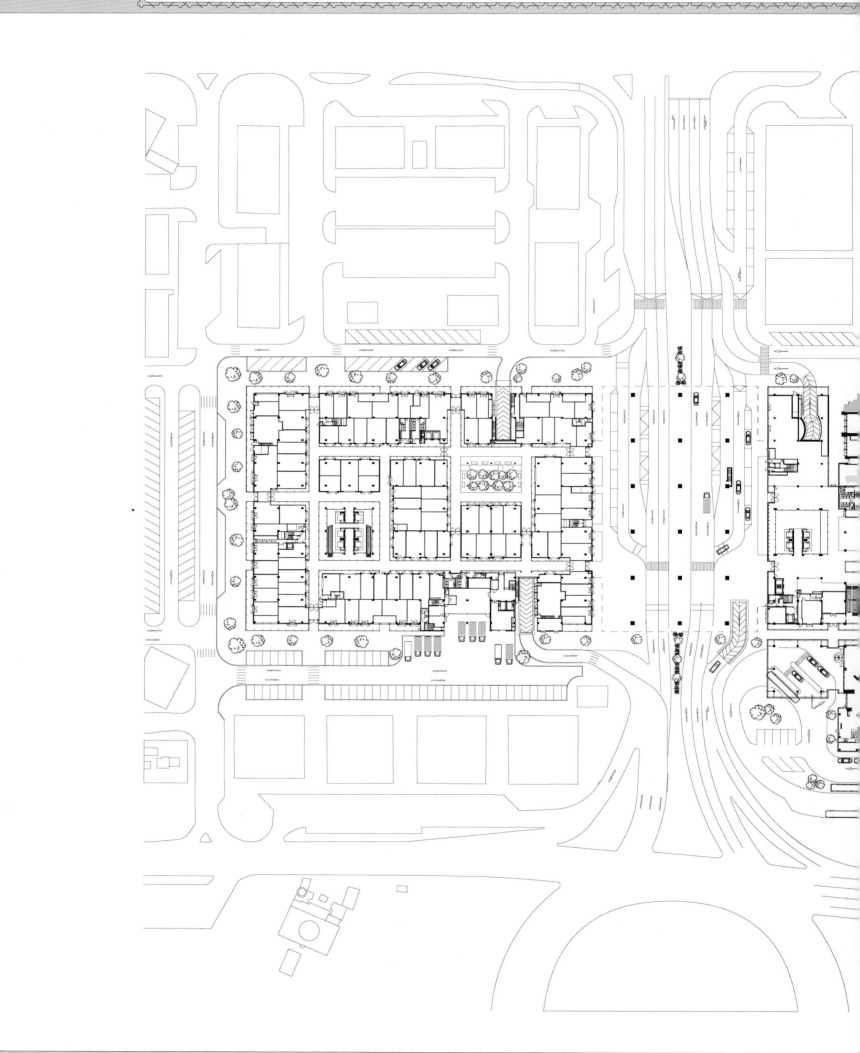

地点：阿拉伯联合酋长国阿布扎比　占地面积：57 100平方米　业主：ALDAR Properties PJSC　建筑设计：福斯特建筑事务所
Location: Abu Dhabi, United Arab Emirates　Site Area: 57,100m²　Client: ALDAR Properties PJSC　Architecture Design: Foster + Partners

阿布扎比阿尔达尔中央市场
Aldar Central Market, Abu Dhabi

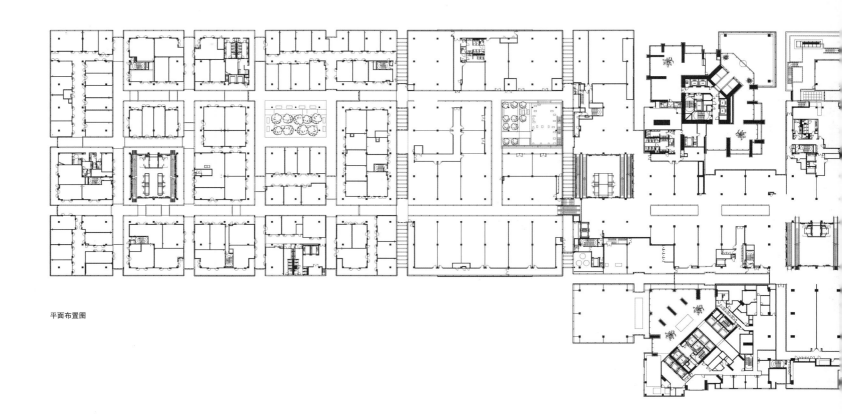

平面布置图

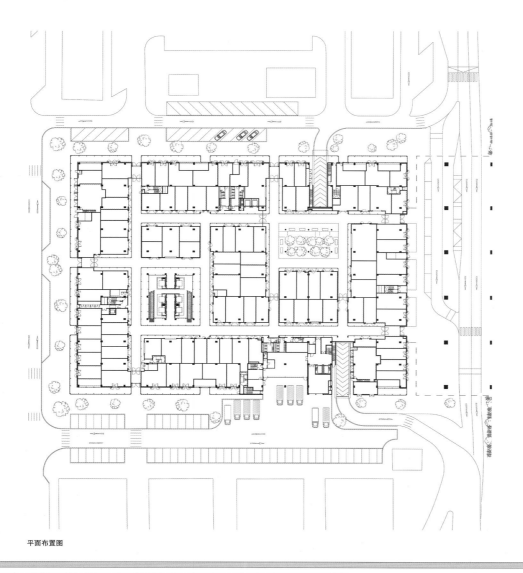

平面布置图

地点: 阿拉伯联合酋长国阿布扎比　占地面积: 57 100平方米　业主: ALDAR Properties PJSC　建筑设计: 福斯特建筑事务所
Location: Abu Dhabi, United Arab Emirates　Site Area: 57,100m²　Client: ALDAR Properties PJSC　Architecture Design: Foster + Partners

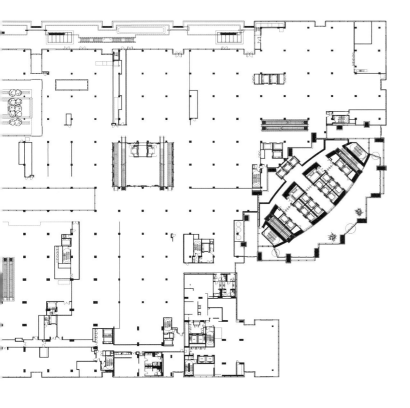

阿布扎比阿尔达尔中央市场
Aldar Central Market, Abu Dhabi

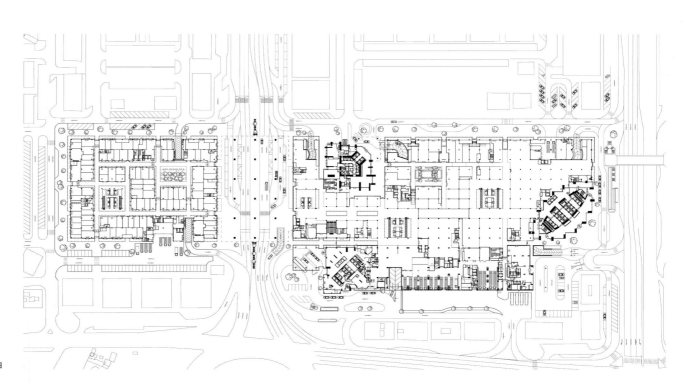

平面布置图

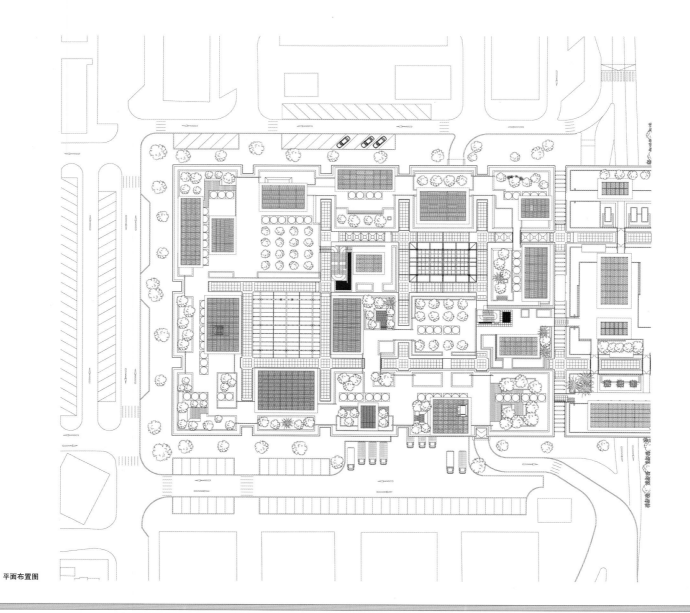

平面布置图

地点：阿拉伯联合酋长国阿布扎比　占地面积：57 100平方米　业主：ALDAR Properties PJSC　建筑设计：福斯特建筑事务所
Location: Abu Dhabi, United Arab Emirates　Site Area: 57,100m²　Client: ALDAR Properties PJSC　Architecture Design: Foster + Partners

阿布扎比阿尔达尔中央市场
Aldar Central Market, Abu Dhabi

地点：阿拉伯联合酋长国阿布扎比　占地面积：57 100平方米　业主：ALDAR Properties PJSC　建筑设计：福斯特建筑事务所
Location: Abu Dhabi, United Arab Emirates   Site Area: 57,100m²   Client: ALDAR Properties PJSC   Architecture Design: Foster + Partners

阿布扎比阿尔达尔中央市场
Aldar Central Market, Abu Dhabi

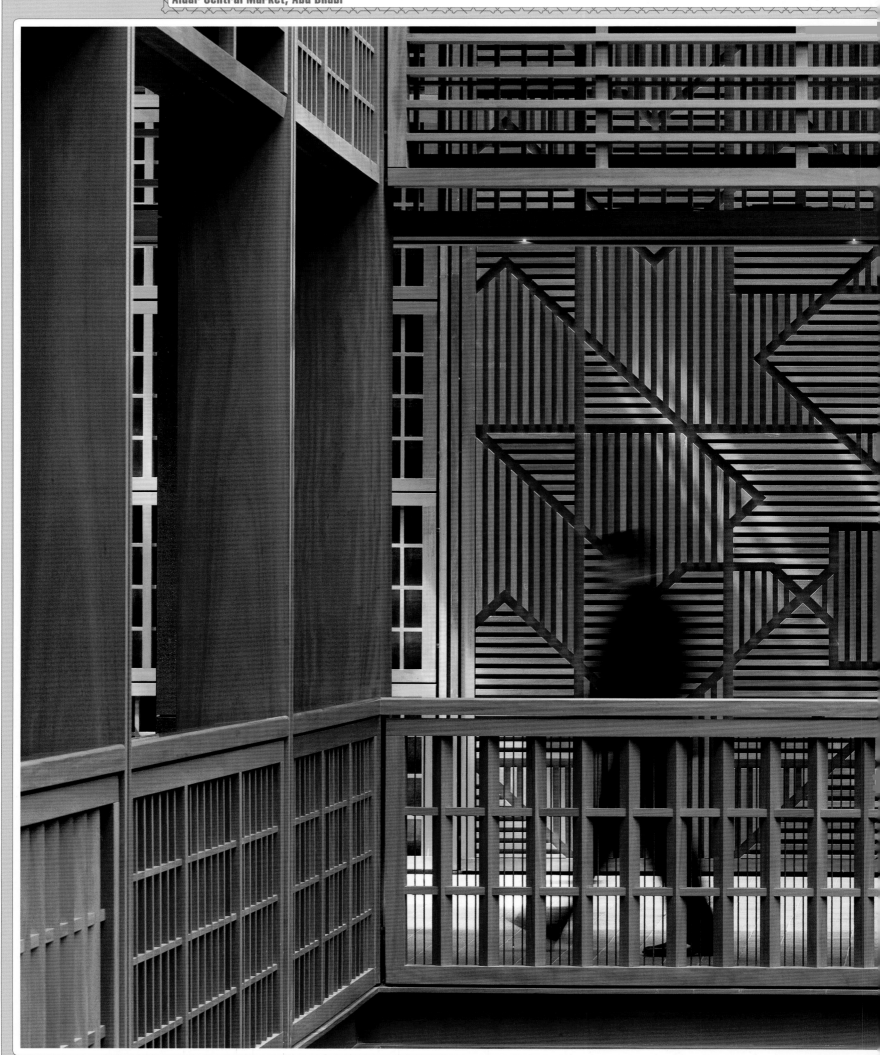

阿布扎比阿尔达尔中央市场
Aldar Central Market, Abu Dhabi

地点：阿拉伯联合酋长国阿布扎比　占地面积: 57 100平方米　业主：ALDAR Properties PJSC　建筑设计：福斯特建筑事务所
Location: Abu Dhabi, United Arab Emirates　Site Area: 57,100m²　Client: ALDAR Properties PJSC　Architecture Design: Foster + Partners

Maras Park
Maras Park

业主：Renaissance Development　设计公司：5+Design　总建筑占地面积：78 000平方米　设计团队：Tim Thurik,
Client: Renaissance Development　Design Company: 5+Design　Gross Building Area: 78,000m²　Design Team: Tim Thurik,

Mark Welz, Mi Sun Lim, Randy Brockman Victoria Brash,　Mohamed ElSheikh, Marshall Ford, Irfan Shaikh, Mark Commentz, Sharon Huang, Warren Riggs, Daniel Penick, Melissa Hsu　地点：土耳其

Mark Welz, Mi Sun Lim, Randy Brockman Victoria Brash,　Mohamed ElSheikh, Marshall Ford, Irfan Shaikh, Mark Commentz, Sharon Huang, Warren Riggs, Daniel Penick, Melissa Hsu　Location: Turkey

Maras Park
Maras Park

业主：Renaissance Development　设计公司：5+Design　总建筑占地面积：78 000平方米　设计团队：Tim Thurik,
Client: Renaissance Development　Design Company: 5+Design　Gross Building Area: 78,000m²　Design Team: Tim Thurik,

北立面图

Mark Welz, Mi Sun Lim, Randy Brockman Victoria Brash,　Mohamed ElSheikh, Marshall Ford, Irfan Shaikh, Mark Commentz, Sharon Huang, Warren Riggs, Daniel Penick, Melissa Hsu　地点：土耳其

Mark Welz, Mi Sun Lim, Randy Brockman Victoria Brash,　Mohamed ElSheikh, Marshall Ford, Irfan Shaikh, Mark Commentz, Sharon Huang, Warren Riggs, Daniel Penick, Melissa Hsu　Location: Turkey

Maras Park
Maras Park

业主：Renaissance Development　设计公司：5+Design　总建筑占地面积：78 000平方米　设计团队：Tim Thurik,
Client: Renaissance Development　Design Company: 5+Design　Gross Building Area: 78,000m²　Design Team: Tim Thurik,

Mark Welz, Mi Sun Lim, Randy Brockman Victoria Brash, Mohamed ElSheikh, Marshall Ford, Irfan Shaikh, Mark Commentz, Sharon Huang, Warren Riggs, Daniel Penick, Melissa Hsu 地点：土耳其
Mark Welz, Mi Sun Lim, Randy Brockman Victoria Brash, Mohamed ElSheikh, Marshall Ford, Irfan Shaikh, Mark Commentz, Sharon Huang, Warren Riggs, Daniel Penick, Melissa Hsu Location: Turkey

Maras Park
Maras Park

业主: Renaissance Development　设计公司: 5+Design　总建筑占地面积: 78 000平方米　设计团队: Tim Thurik,
Client: Renaissance Development　Design Company: 5+Design　Gross Building Area: 78,000m²　Design Team: Tim Thurik,

剖面图

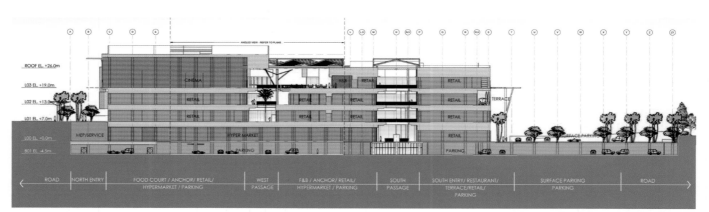

剖面图

Mark Welz, Mi Sun Lim, Randy Brockman Victoria Brash, Mohamed ElSheikh, Marshall Ford, Irfan Shaikh, Mark Commentz, Sharon Huang, Warren Riggs, Daniel Penick, Melissa Hsu   地点：土耳其
Mark Welz, Mi Sun Lim, Randy Brockman Victoria Brash, Mohamed ElSheikh, Marshall Ford, Irfan Shaikh, Mark Commentz, Sharon Huang, Warren Riggs, Daniel Penick, Melissa Hsu   Location: Turkey

Maras Park
Maras Park

业主：Renaissance Development　设计公司：5+Design　总建筑占地面积：78 000平方米　设计团队：Tim Thurik,
Client: Renaissance Development　Design Company: 5+Design　Gross Building Area: 78,000m²　Design Team: Tim Thurik,

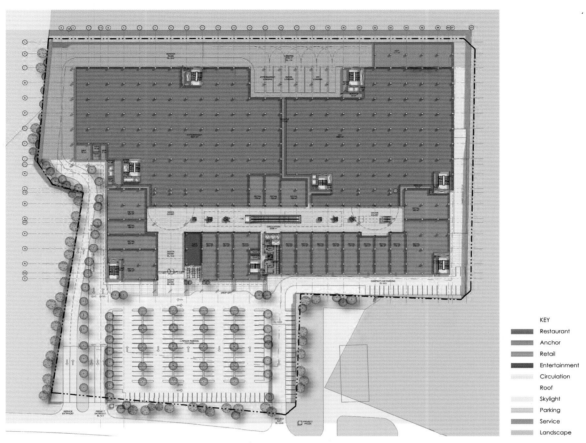

KEY
Restaurant
Anchor
Retail
Entertainment
Circulation
Roof
Skylight
Parking
Service
Landscape

平面布置图

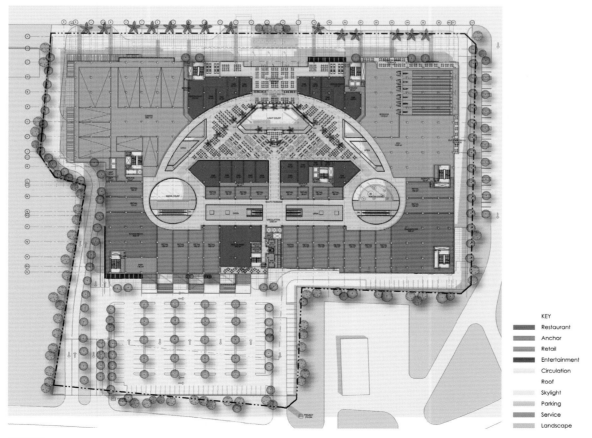

KEY
Restaurant
Anchor
Retail
Entertainment
Circulation
Roof
Skylight
Parking
Service
Landscape

平面布置图

地点：韩国大邱市　占地面积：50 000平方米　总可租面积：45 000平方米　开发商：彩色广场　设计公司：美国捷得国际建筑师事务所
Location: Daegu, South Korea　Site Area: 50,000m²　Gross Leasable Area: 45,000m²　Developer: Color Square　Design Company: The Jerde Partnership

TENNIS COURTS

SWIMMING POOL

GRAPHIC SCALE

TRUE NORTH　PLAN NORTH

0　　75　　125　　250M

在一个靠近大邱世界杯体育馆的地下车库改造招标中，受到当地纺织工业的启发，设计师提交了一个高度标志性的马赛克表皮形式的概念方案并被采纳。设计更从大邱世界杯体育馆象征全球团结获得灵感，马赛克表皮效果暗示了来访问这个集娱乐、展览、餐饮、体育活动于一体的新中心对于世界不同文化的交织与融合。

为进一步满足当地居民需求，项目被设计成可持续发展的文化和艺术品收藏的社区，通过生活花园、屋顶公园、水景和其他景观将场地变成城市绿洲。项目不仅将场地变成一个集运动与景观于一体的综合体，保证连续性与公共性，而且使其成为一个文化活动的休闲中心，有序地将人们从世界杯体育馆引导至其他体育馆和未来的棒球场。

项目拥有50 000平方米惊人的零售娱乐空间，将会加强这一国际运动场地的公共体验。

受到体育馆标志性的曲线屋顶象征全球团结的启发，项目外墙覆盖了独特绚丽的马赛克图案，暗示世界多元文化。这种独特的马赛克建筑外墙风格也是受到当地著名的纺织工业和丰富多彩编织物的影响。

这些新的外墙和地下两层零售娱乐空间围绕体育馆的停车空间螺旋上升，形成一系列人行景观步道，并抬高了公园的位置，创造了连续的外部和内部路径，连接了零售娱乐空间与周边体育场地，包括一个次要的体育馆与未来的棒球馆。

综合体的中心是一个水景花园，户外下沉式广场与水景提供了新的社交场地与活动场地。商店、餐厅和娱乐场所围绕水景花园布置，放射状的小巷和玻璃幕墙为多个地下区域提供了自然采光。备受期待的娱乐区设置了一个1 400个座位的音乐会活动场所和六个电影院。

项目中大量采用了自然有机元素，提高了公园的质量，为游客创造了一片绿洲。中心的可持续策略包括中水灌溉植物、地源热泵及可回收材料。

大邱彩色广场购物中心
Color Square Stadium Mall Daegu

地点：韩国大邱市　占地面积：50 000平方米　总可租面积：45 000平方米　开发商：彩色广场　设计公司：美国捷得国际建筑师事务所
Location: Daegu, South Korea　Site Area: 50,000m²　Gross Leasable Area: 45,000m²　Developer: Color Square　Design Company: The Jerde Partnership

大邱彩色广场购物中心
Color Square Stadium Mall Daegu

地点：韩国大邱市　占地面积：50 000平方米　总可租面积：45 000平方米　开发商：彩色广场　设计公司：美国捷得国际建筑师事务所
Location: Daegu, South Korea　Site Area: 50,000m²　Gross Leasable Area: 45,000m²　Developer: Color Square　Design Company: The Jerde Partnership

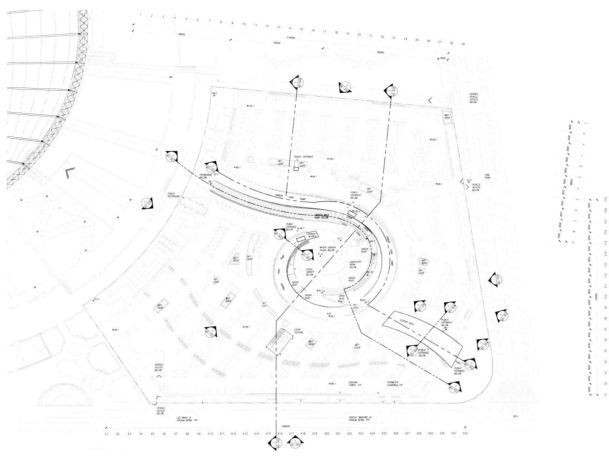

平面布置图

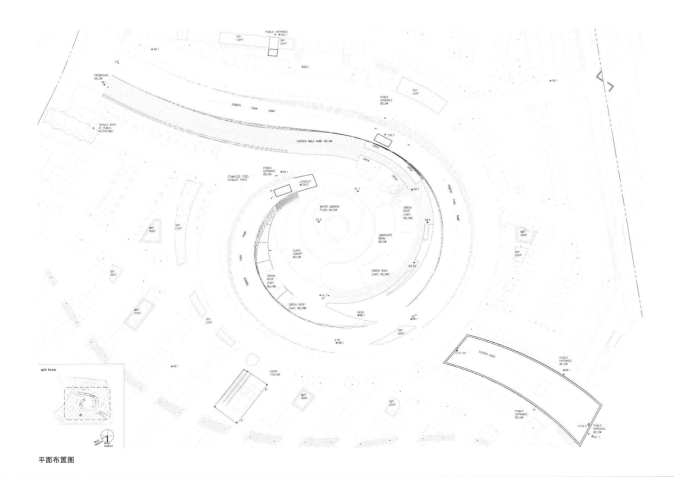

平面布置图

大邱彩色广场购物中心
Color Square Stadium Mall Daegu

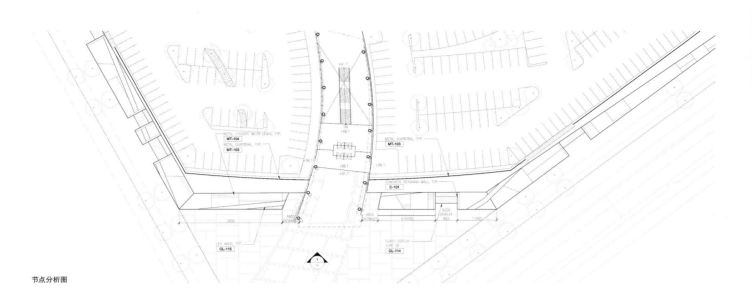

节点分析图

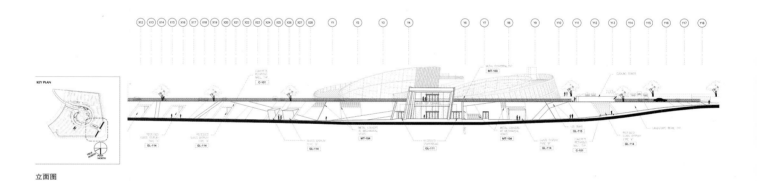

立面图

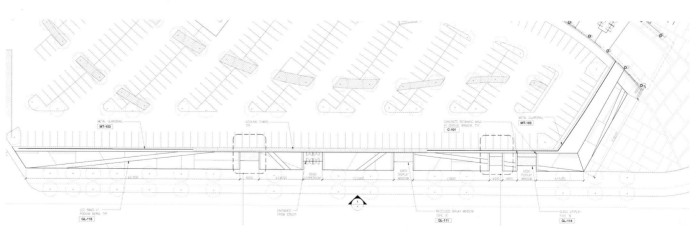

节点分析图

地点：韩国大邱市　占地面积：50 000平方米　总可租面积：45 000平方米　开发商：彩色广场　设计公司：美国捷得国际建筑师事务所
Location: Daegu, South Korea　Site Area: 50,000m²　Gross Leasable Area: 45,000m²　Developer: Color Square　Design Company: The Jerde Partnership

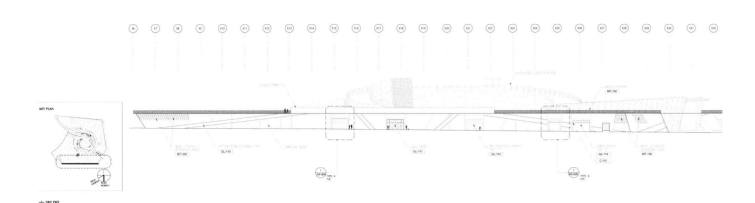

立面图

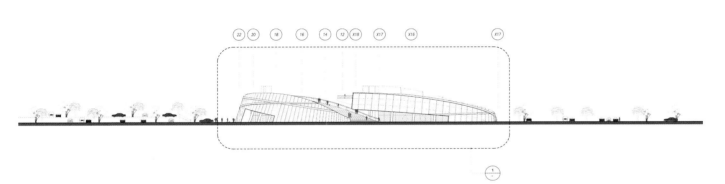

立面图

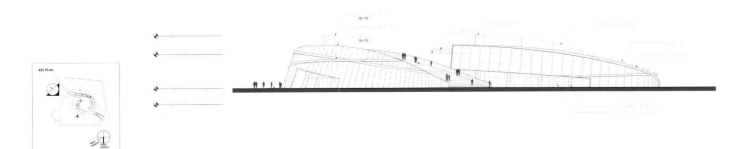

立面图

大邱彩色广场购物中心
Color Square Stadium Mall Daegu

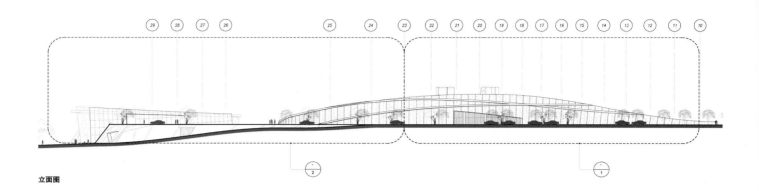

立面图

立面图

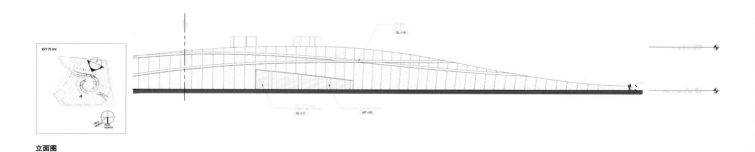

立面图

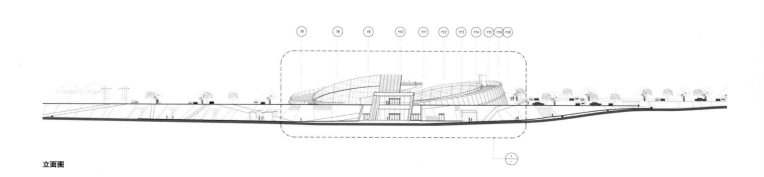

立面图

地点：韩国大邱市 占地面积：50 000平方米 总可租面积：45 000平方米 开发商：彩色广场 设计公司：美国捷得国际建筑师事务所
Location: Daegu, South Korea   Site Area: 50,000m²   Gross Leasable Area: 45,000m²   Developer: Color Square   Design Company: The Jerde Partnership

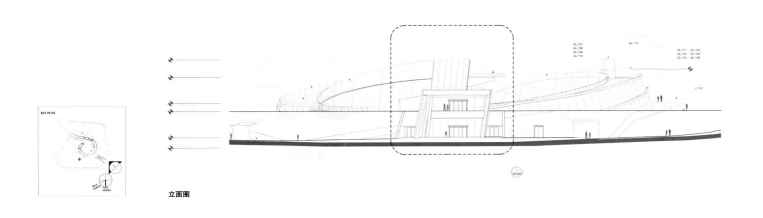

立面图

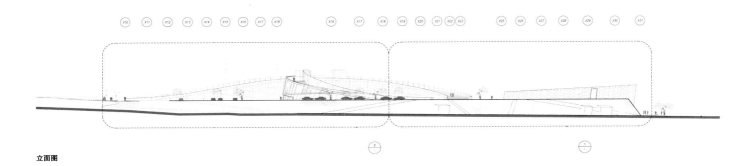

立面图

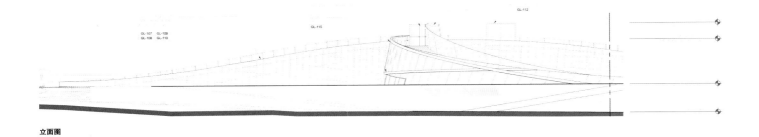

立面图

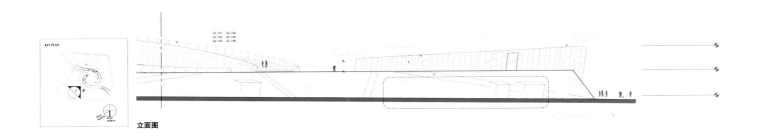

立面图

大邱彩色广场购物中心
Color Square Stadium Mall Daegu

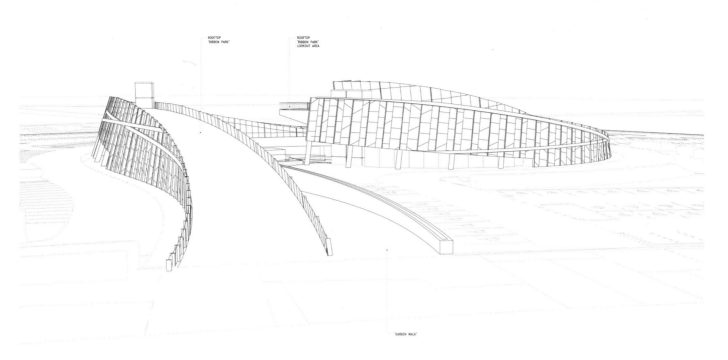

ROOFTOP "RIBBON PARK"

ROOFTOP "RIBBON PARK" LOOKOUT AREA

'GARDEN WALK'

节点分析图

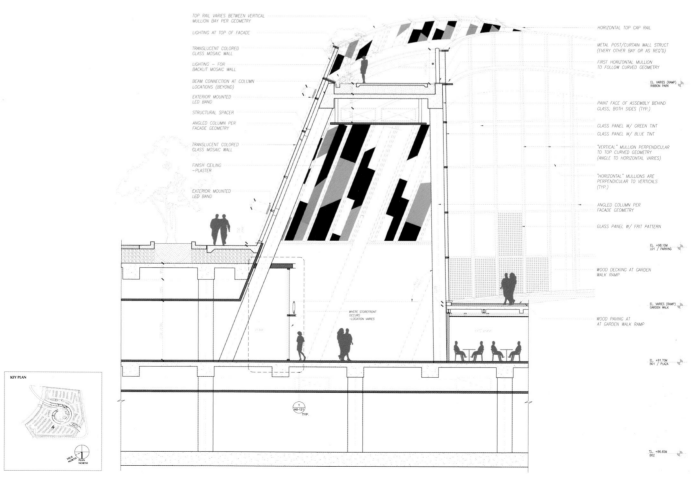

TOP RAIL VARIES BETWEEN VERTICAL MULLION BAY PER GEOMETRY

LIGHTING AT TOP OF FACADE

TRANSLUCENT COLORED GLASS MOSAIC WALL

LIGHTING - FOR BACKLIT MOSAIC WALL

BEAM CONNECTION AT COLUMN LOCATIONS (BEYOND)

EXTERIOR MOUNTED LED BAND

STRUCTURAL SPACER

ANGLED COLUMN PER FACADE GEOMETRY

TRANSLUCENT COLORED GLASS MOSAIC WALL

FINISH CEILING - PLASTER

EXTERIOR MOUNTED LED BAND

HORIZONTAL TOP CAP RAIL

METAL POST/CURTAIN WALL STRUCT (EVERY OTHER BAY OR AS REQ'D)

FIRST HORIZONTAL MULLION TO FOLLOW CURVED GEOMETRY

EL. VARIES (RAMP) RIBBON PARK

PAINT FACE OF ASSEMBLY BEHIND GLASS, BOTH SIDES (TYP.)

GLASS PANEL W/ GREEN TINT

GLASS PANEL W/ BLUE TINT

"VERTICAL" MULLION PERPENDICULAR TO TOP CURVED GEOMETRY (ANGLE TO HORIZONTAL VARIES)

"HORIZONTAL" MULLIONS ARE PERPENDICULAR TO VERTICALS (TYP.)

ANGLED COLUMN PER FACADE GEOMETRY

GLASS PANEL W/ FRIT PATTERN

EL. +98.10M L01 / PARKING

WOOD DECKING AT GARDEN WALK RAMP

EL. VARIES (RAMP) GARDEN WALK

WOOD PAVING AT AT GARDEN WALK RAMP

EL. +91.70M B01 / PLAZA

WHERE STOREFRONT OCCURS -LOCATION VARIES

EL. +86.65M B02

KEY PLAN

NORTH

墙体剖面分析图

地点：韩国大邱市　占地面积：50 000平方米　总可租面积：45 000平方米　开发商：彩色广场　设计公司：美国捷得国际建筑师事务所
Location: Daegu, South Korea　Site Area: 50,000m²　Gross Leasable Area: 45,000m²　Developer: Color Square　Design Company: The Jerde Partnership

大邱彩色广场购物中心
Color Square Stadium Mall Daegu

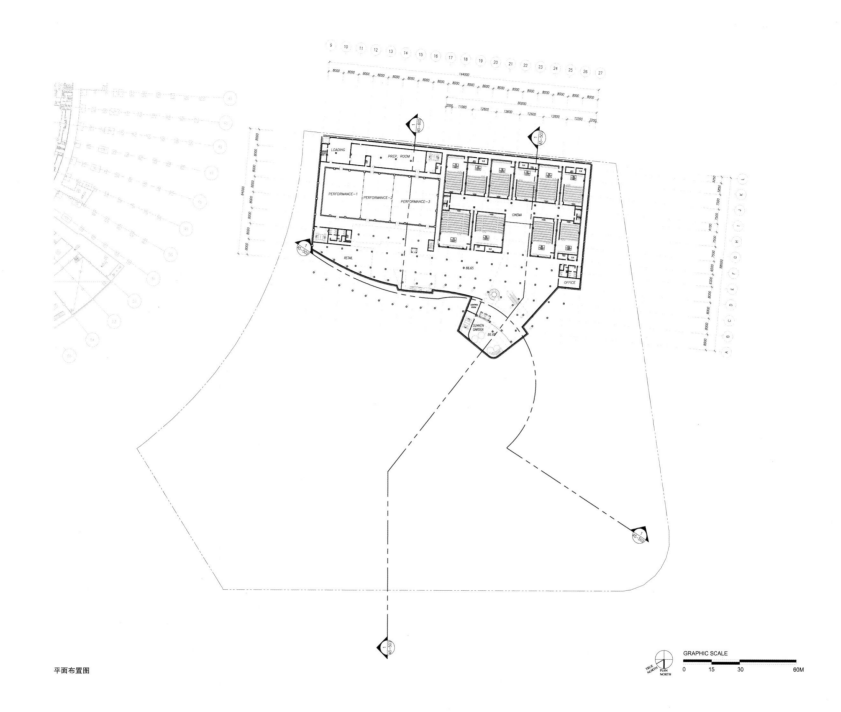

平面布置图

GRAPHIC SCALE

地点：韩国大邱市　占地面积：50 000平方米　总可租面积：45 000平方米　开发商：彩色广场　设计公司：美国捷得国际建筑师事务所
Location: Daegu, South Korea　Site Area: 50,000m²　Gross Leasable Area: 45,000m²　Developer: Color Square　Design Company: The Jerde Partnership

商业街＆社区商业

Commercial Streets & Community Commerce

南京夫子庙改造项目（一期）
The Confucian Temple Renovation Phase I, Nanjing

业主：夫子庙管委会　设计单位：DC国际建筑设计事务所　设计师：董屹、程久军、赵维、金渊　用地面积：11 000平方米　建筑面积：12 000平方米　竣工时间：2014年1月　地点：南京市秦淮区贡院街
Client: The Administrative Committee of the Confucian Temple　Design Company: DC Alliance Architecture Design　Designers: Dong Yi, Cheng Jiujun, Zhao Wei, Jin Yuan　Site Area: 11,000m²　Building Area: 12,000m²
Completion Time: Jan. 2014　Location: Gongyuan Street, Qinhuai District, Nanjing, Jiangsu Province, China

南京夫子庙改造项目（一期）
The Confucian Temple Renovation Phase I, Nanjing

南京夫子庙改造项目（一期）
The Confucian Temple Renovation Phase I, Nanjing

总平面图

业主：夫子庙管委会　设计单位：DC国际建筑设计事务所　设计师：董屹、程久军、赵维、金渊　用地面积：11 000平方米　建筑面积：12 000平方米　竣工时间：2014年1月　地点：南京市秦淮区贡院街
Client: The Administrative Committee of the Confucian Temple　Design Company: DC Alliance Architecture Design　Designers: Dong Yi, Cheng Jiujun, Zhao Wei, Jin Yuan　Site Area: 11,000m²　Building Area: 12,000m²
Completion Time: Jan. 2014　Location: Gongyuan Street, Qinhuai District, Nanjing, Jiangsu Province, China

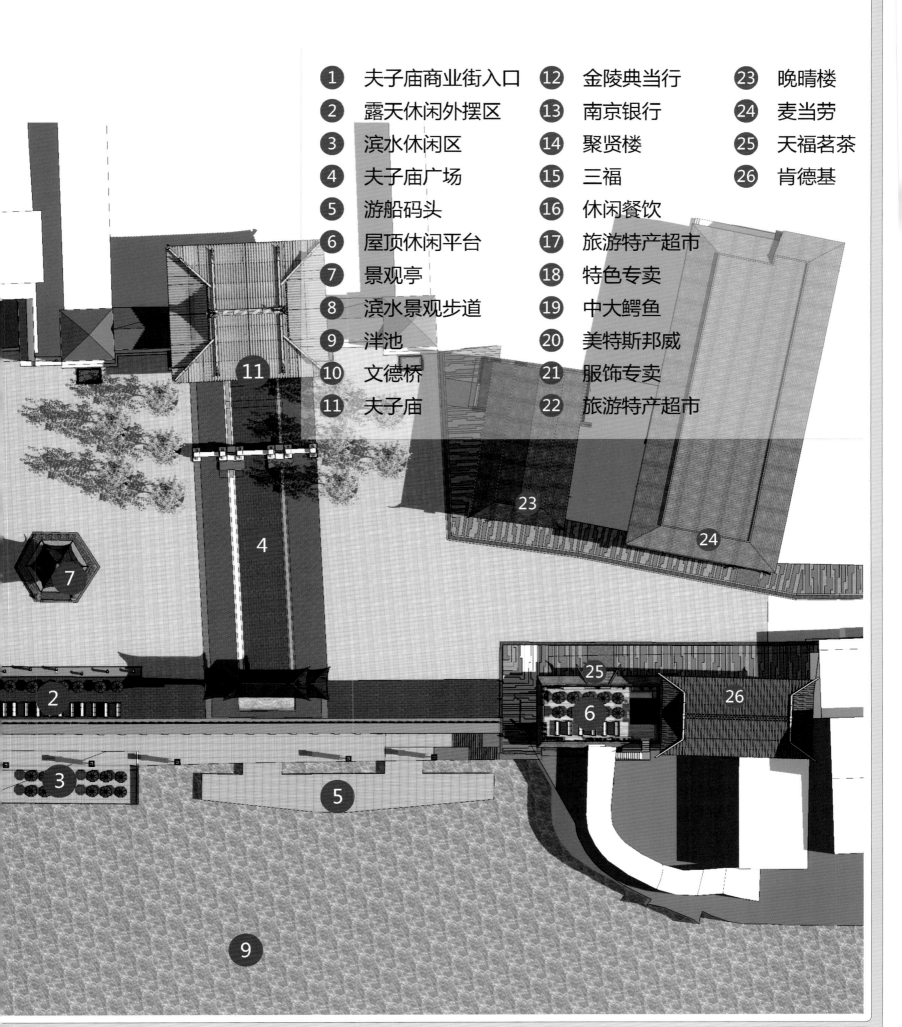

1 夫子庙商业街入口
2 露天休闲外摆区
3 滨水休闲区
4 夫子庙广场
5 游船码头
6 屋顶休闲平台
7 景观亭
8 滨水景观步道
9 泮池
10 文德桥
11 夫子庙

12 金陵典当行
13 南京银行
14 聚贤楼
15 三福
16 休闲餐饮
17 旅游特产超市
18 特色专卖
19 中大鳄鱼
20 美特斯邦威
21 服饰专卖
22 旅游特产超市

23 晚晴楼
24 麦当劳
25 天福茗茶
26 肯德基

南京夫子庙改造项目（一期）
The Confucian Temple Renovation Phase I, Nanjing

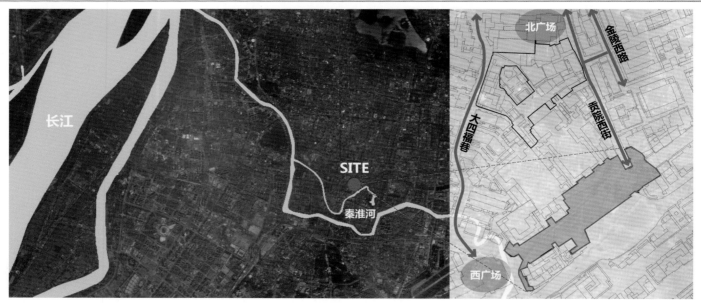

区位图

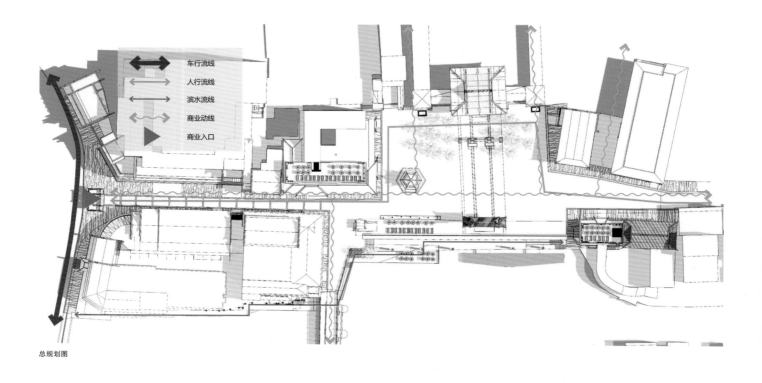

总规划图

南京夫子庙改造项目(一期)位于南京市秦淮区贡院街从西牌坊到贡院西街的整个街道和相关衍生区域,以及贡院西街的局部区域。项目由秦淮区政府牵头,夫子庙管理会具体操作,时尚创建为顾问公司,DC国际为设计单位。公共区域的改造结合雨污分流工程同步实施。项目采取政府统筹、房屋业主出资建造、设计公司和顾问公司对整体效果把关的方式实施。

夫子庙改造前,建筑多为20世纪80年代按明清时期建筑形式修建的,在经过30多年风雨洗礼后立面破坏严重,且很多立面形式与现代功能冲突。在夫子庙现状房屋的产权与租赁关系上,现单一租户对应一个产权人,而一个产权人往往对应多个租户。这种租赁关系比较容易在单栋建筑上取得形态上的变化。在立面现状上存在的问题也是由于这种租赁关系,没有统一且具指向性的设计规划,导致现状的立面体块关系复杂。一层商铺整体视觉凌乱,缺乏控制,二层立面做法对原有传统古建的改动欠妥。故本次设计的目的是恢复夫子庙片区传统明清建筑的风貌,根据产权与业态梳理整个街区建筑群的立面关系,同时使建筑立面尽可能满足文化旅游商业的需求。

在保持现状空间布局的情况下,我们在整体设计策略上提出"两轴四点"的空间分区。"两轴"即"文化礼仪"轴线与"商业体验"轴线,"四点"即城市入口、文德桥头、天下文枢、两街交汇,以营造出"整体的传统氛围,时尚的商业体验"的气氛。

在建筑的设计改造上,我们遵循"一层时尚,二层传统"的原则,保持城市尺度的传统空间体验与近人尺度的时尚商业感受。不同的单栋建筑结合处,设计师采用现代设计思路,使其传统商业街中成为点睛之笔。

在材料选用上,选用现代材料来表达传统建筑的韵味,并与现代建筑的质感的结合。在建筑色调的选用上采用红、白、灰三种色调,红为窗,白为墙,灰为瓦,金为点缀,以在整体建筑上进行统一,对粉墙黛瓦的江南建筑符号进行了新的诠释。

考虑建筑立面的适应性与长久性,设计师在设计中充分考虑建筑的建筑要素和商铺要素,将建筑要素和商铺要素分离,二层以上基本为建筑要素,一层的建筑要素主要为立柱、门头和相关展示墙面,其余为店铺要素。在前期设计中根据已有的建筑业态定义出几种类型的店铺,并根据业态布局来组织整体立面建筑要

业主：夫子庙管委会　设计单位：DC国际建筑设计事务所　设计师：董屹、程久军、赵维、金渊　用地面积：11 000平方米　建筑面积：12 000平方米　竣工时间：2014年1月　地点：南京市秦淮区贡院街
Client: The Administrative Committee of the Confucian Temple　Design Company: DC Alliance Architecture Design　Designers: Dong Yi, Cheng Jiujun, Zhao Wei, Jin Yuan　Site Area: 11,000m²　Building Area: 12,000m²
Completion Time: Jan. 2014　Location: Gongyuan Street, Qinhuai District, Nanjing, Jiangsu Province, China

素。在后期，各商铺业主根据实际需求自行进行调整，增强各商铺的灵活性、多样性。而建筑要素是在一段时间内控制不变的，以此在保证商业多样性的同时保持传统建筑的风貌。

这个项目在方案阶段有过许许多多好的精彩的想法，但在实际操作中，设计师发现这些房子本身的破坏过于严重，很多现场的设计的发挥较多，策划公司对于业态更替的想法也由于房屋产权的复杂关系而没有全部完成，这样一来对于一层店铺立面更新的想法其实是需要一个循序渐进的过程来完成的，整体的业态和立面氛围也需要一个相当长的时间来提升。设计师认为对于南京夫子庙改造（一期）的设计我们做的只是一个框架，还需要很多内容填充进来，一种自发生长的、多方参与的文化街区才能真正融入人们的生活。

南京夫子庙改造项目（一期）
The Confucian Temple Renovation Phase I, Nanjing

业主：夫子庙管委会　设计单位：DC国际建筑设计事务所　设计师：董屹、程久军、赵维、金渊　用地面积：11 000平方米　建筑面积：12 000平方米　竣工时间：2014年1月　地点：南京市秦淮区贡院街
Client: The Administrative Committee of the Confucian Temple　Design Company: DC Alliance Architecture Design　Designers: Dong Yi, Cheng Jiujun, Zhao Wei, Jin Yuan　Site Area: 11,000m²　Building Area: 12,000m²
Completion Time: Jan. 2014　Location: Gongyuan Street, Qinhuai District, Nanjing, Jiangsu Province, China

南京夫子庙改造项目（一期）
The Confucian Temple Renovation Phase I, Nanjing

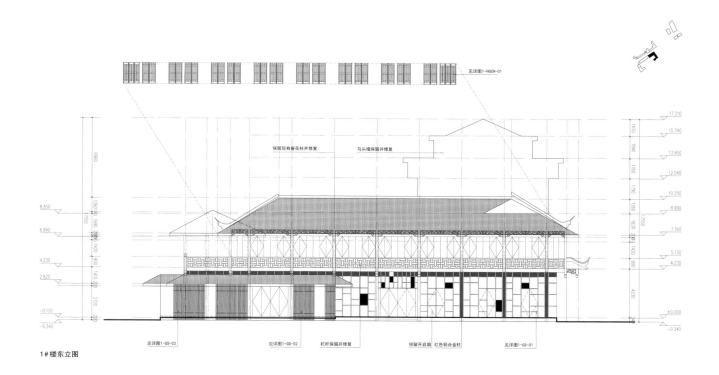

保留现有番花柱并修复　　马头墙保留并修复

见详图1-HGQH-01

见详图1-GS-02　　　见详图1-GS-02　栏杆保留并修复　　预留开启扇　红色铝合金柱　见详图1-GS-01

1#楼东立图

业主: 夫子庙管委会   设计单位: DC国际建筑设计事务所   设计师: 董屹、程久军、赵维、金渊   用地面积: 11 000平方米   建筑面积: 12 000平方米   竣工时间: 2014年1月   地点: 南京市秦淮区贡院街
Client: The Administrative Committee of the Confucian Temple   Design Company: DC Alliance Architecture Design   Designers: Dong Yi, Cheng Jiujun, Zhao Wei, Jin Yuan   Site Area: 11,000m²   Building Area: 12,000m²
Completion Time: Jan. 2014   Location: Gongyuan Street, Qinhuai District, Nanjing, Jiangsu Province, China

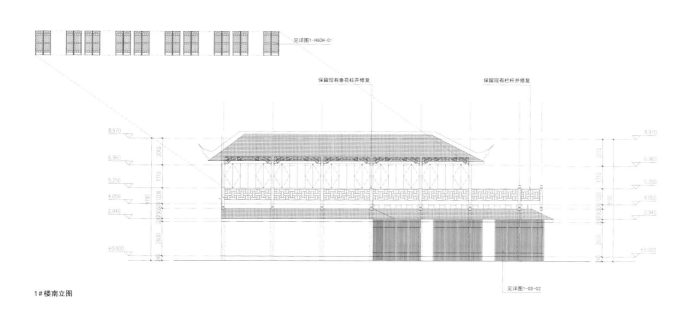

1#楼南立图

南京夫子庙改造项目（一期）
The Confucian Temple Renovation Phase I, Nanjing

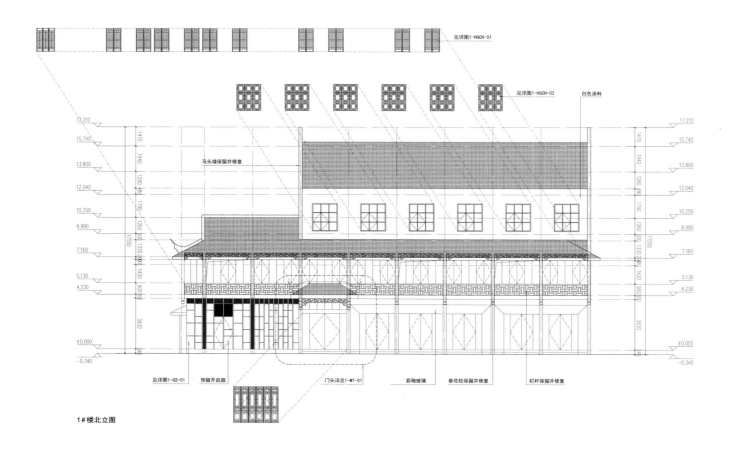

见详图1-HGCH-01

见详图1-HGCH-02

白色涂料

马头墙保留并修复

见详图1-GS-01　　预留开启扇　　　门头详见1-MT-01　　彩釉玻璃　　垂花柱保留并修复　　栏杆保留并修复

1#楼北立图

业主: 夫子庙管委会　设计单位: DC国际建筑设计事务所　设计师: **董屹、程久军、赵维、金渊**　用地面积: **11 000平方米**　建筑面积: **12 000平方米**　竣工时间: **2014年1月**　地点: 南京市秦淮区贡院街
Client: The Administrative Committee of the Confucian Temple　Design Company: DC Alliance Architecture Design　Designers: Dong Yi, Cheng Jiujun, Zhao Wei, Jin Yuan　Site Area: 11,000m²　Building Area: 12,000m²
Completion Time: Jan. 2014　Location: Gongyuan Street, Qinhuai District, Nanjing, Jiangsu Province, China

南京夫子庙改造项目（一期）
The Confucian Temple Renovation Phase I, Nanjing

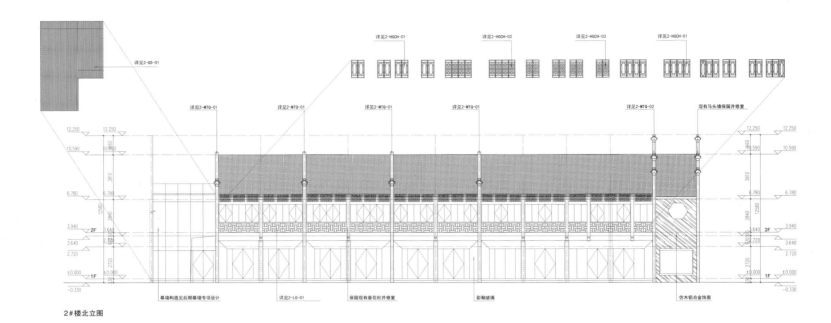

详见2-GS-01

详见2-HGCH-01   详见2-HGCH-02   详见2-HGCH-03   详见2-HGCH-01

详见2-MTQ-01   详见2-MTQ-01   详见2-MTQ-01   详见2-MTQ-01   详见2-MTQ-02   现有马头墙保留并修复

幕墙构造见后期幕墙专项设计   详见2-LG-01   保留现有垂花柱并修复   彩釉玻璃   仿木铝合金饰面

2#楼北立图

业主：夫子庙管委会　设计单位：DC国际建筑设计事务所　设计师：董屹、程久军、赵维、金渊　用地面积：11 000平方米　建筑面积：12 000平方米　竣工时间：2014年1月　地点：南京市秦淮区贡院街
Client: The Administrative Committee of the Confucian Temple　Design Company: DC Alliance Architecture Design　Designers: Dong Yi, Cheng Jiujun, Zhao Wei, Jin Yuan　Site Area: 11,000m²　Building Area: 12,000m²
Completion Time: Jan. 2014　Location: Gongyuan Street, Qinhuai District, Nanjing, Jiangsu Province, China

南京夫子庙改造项目（一期）
The Confucian Temple Renovation Phase I, Nanjing

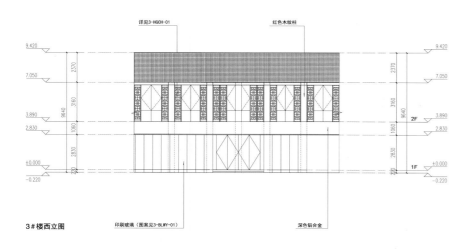

3#楼西立图

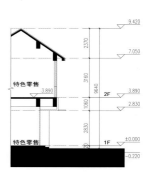

3#楼1-1剖面图

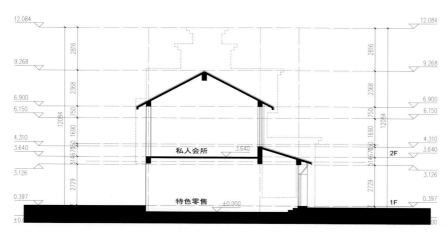

4#楼1-1剖面图

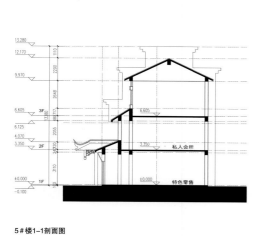

5#楼1-1剖面图

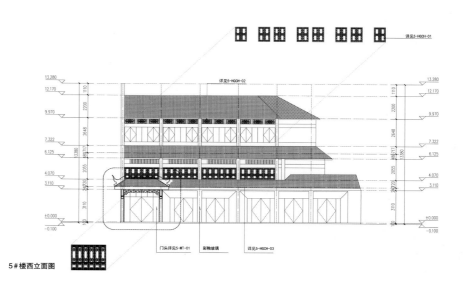

5#楼西立面图

业主：夫子庙管委会　设计单位：DC国际建筑设计事务所　设计师：董屹、程久军、赵维、金渊　用地面积：11 000平方米　建筑面积：12 000平方米　竣工时间：2014年1月　地点：南京市秦淮区贡院街
Client: The Administrative Committee of the Confucian Temple　Design Company: DC Alliance Architecture Design　Designers: Dong Yi, Cheng Jiujun, Zhao Wei, Jin Yuan　Site Area: 11,000m²　Building Area: 12,000m²
Completion Time: Jan. 2014　Location: Gongyuan Street, Qinhuai District, Nanjing, Jiangsu Province, China

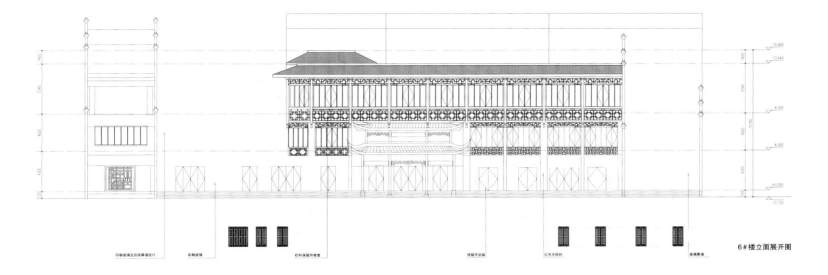

6#楼立面展开图

Located in Gongyuan Street, Qinhuai District, Nanjing, the project covers the whole section from Xipaifang to Gongyuan West Street, as well as relevant derivative area and partial area of Gongyuan West Street. The project is leaded by the government of Qinhuai District and undertaken by the Administrative Committee of the Confucian Temple. Lifestyle Creating was invited as the consultant and DC Alliance as the design company. The public area renovation was synchronized with the rain-sewage division. The project was carried out through governmental coordination, property owners' investment, design company and consultant company's overall control. Before renovation, the Confucian Temple featured buildings built during 1980s in the style of Ming and Qing Dynasties. Having been exposed to rain and wind for more than 30 years, the buildings' facades are heavily damaged, and meanwhile, most of them conflict with their modern function. As for the ownership and lease of the existing properties, one tenant had only one property owner, while one property owner had several tenants. This enabled a single building to have variation on its facade. However, this also caused problems to the facade, due to the lack of unified and directive design and plan. The existing facades were quite complex. The facade on the first floor looked quite disorder and out of control, while the facade on the second floor did improper alteration to the original traditional ancient building. Therefore, the renovation design is aimed to recover the traditional architectural style of Ming and Qing Dynasties in the Confucian Temple area; clarify the buildings' facades in the area according to property right and function; enable the facade to meet culture, tourism and commerce functions.

While remaining the existing space layout, DC Alliance proposed "two axes, four points" space division. Two axes refer to cultural etiquette and commercial experience. Four points stand for city entrance, Wende Bridge head, Tianxiawenshu, intersection of two streets. As a result, an ambiance of overall traditional atmosphere and fashionable commercial experience is created.

In the renovation design of the buildings, DC Alliance follows the principle of fashionable first floor and traditional second floor, realizing urbanized traditional space experience and humanized fashionable commercial touch. At the joint of different buildings, modern design concept is introduced as a finishing touch that stands out in the traditional commercial street.

As for the material application, modern materials are chosen to express the charm of traditional building in combination with the quality of modern building. As for the hue, red, white and gray are adopted— red window, white wall and gray tile. Besides, gold is introduced as embellishment, bringing new interpretation to the typical architectural symbol in the regions south of the Yangtze River.

In view of the facade's adaptability and durability, DC Alliance paid great attention to architectural factor and store factor, which are apart from each other. Architectural factor is dominant above second floor, while on first floor, store factor is dominant, except for architectural factors like columns, store fronts and displaying walls. During the early designing, several store functions are defined based on the existing building function, and overall facade is determined by the function layout, with unified architectural factor. During the later period, store factor can be adjusted to meet the store owner's requirements, thus to add flexibility and variety to his store. But architectural factor is controlled to remain the same within a certain period. The result is not only commercial diversity is ensured, but also traditional architecture is remained.

During the proposal stage, the project was infused with many wonderful ideas. However, in actual operation, DC Alliance found that these buildings were over damaged, which then resulted in more improvisations rather than fulfilling those wonderful ideas. Besides, the planning company's idea of changing function was not completed satisfactorily as well. In this way, DC Alliance's idea of renovating the store facade of first floor requires a gradual process to be realized. The overall function and facade atmosphere also need a quite long time to upgrade. DC Alliance maintains that the design for the project is only a frame that needs to be filled with contents. Only a spontaneously grown and diversified cultural street can truly integrate into people's life.

南京夫子庙改造项目（一期）
The Confucian Temple Renovation Phase I, Nanjing

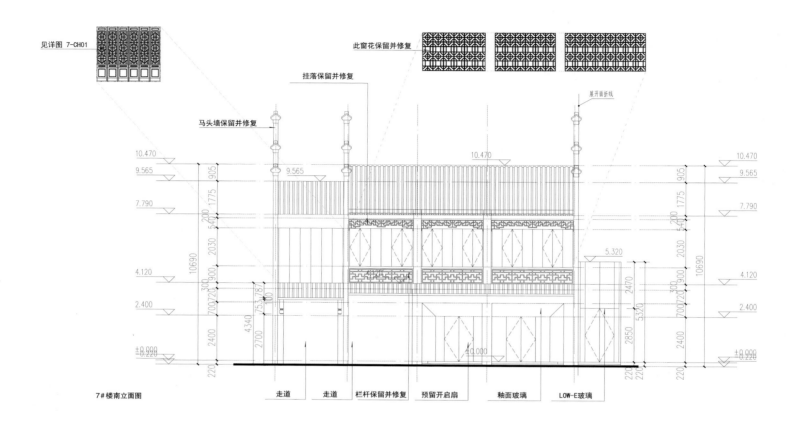

见详图 7-CH01

此窗花保留并修复

挂落保留并修复

展开面折线

马头墙保留并修复

7#楼南立面图

走道　走道　栏杆保留并修复　预留开启扇　釉面玻璃　LOW-E玻璃

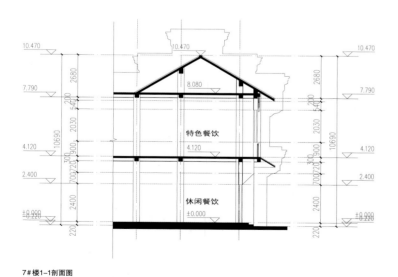

特色餐饮

休闲餐饮

7#楼1-1剖面图

走道

7#楼2-2剖面图

业主：夫子庙管委会　设计单位：DC国际建筑设计事务所　设计师：董屹、程久军、赵维、金渊　用地面积：11 000平方米　建筑面积：12 000平方米　竣工时间：2014年1月　地点：南京市秦淮区贡院街
Client: The Administrative Committee of the Confucian Temple　Design Company: DC Alliance Architecture Design　Designers: Dong Yi, Cheng Jiujun, Zhao Wei, Jin Yuan　Site Area: 11,000m²　Building Area: 12,000m²
Completion Time: Jan. 2014　Location: Gongyuan Street, Qinhuai District, Nanjing, Jiangsu Province, China

贡院街东段6改造后

贡院街东段6改造前

贡院西街2改造后

贡院西街2改造前

南京夫子庙改造项目（一期）
The Confucian Temple Renovation Phase I, Nanjing

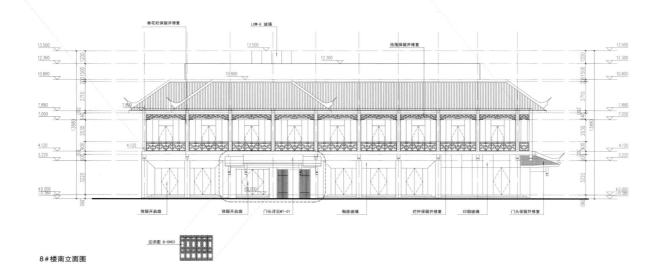

8#楼南立面图

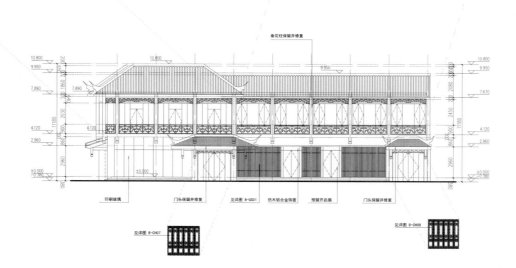

8#楼东立面图

业主：夫子庙管委会　设计单位：DC国际建筑设计事务所　设计师：董屹、程久军、赵维、金渊　用地面积：11 000平方米　建筑面积：12 000平方米　竣工时间：2014年1月　地点：南京市秦淮区贡院街
Client: The Administrative Committee of the Confucian Temple　Design Company: DC Alliance Architecture Design　Designers: Dong Yi, Cheng Jiujun, Zhao Wei, Jin Yuan　Site Area: 11,000m²　Building Area: 12,000m²
Completion Time: Jan. 2014　Location: Gongyuan Street, Qinhuai District, Nanjing, Jiangsu Province, China

状元及第2改造后

状元及第2改造前

状元及第1改造后

状元及第1改造前

南京夫子庙改造项目（一期）
The Confucian Temple Renovation Phase I, Nanjing

贡院街东段4改造后

贡院街东段4改造前

贡院街东段3改造后

贡院街东段3改造前

业主：夫子庙管委会　设计单位：DC国际建筑设计事务所　设计师：董屹、程久军、赵维、金渊　用地面积：11 000平方米　建筑面积：12 000平方米　竣工时间：2014年1月　地点：南京市秦淮区贡院街
Client: The Administrative Committee of the Confucian Temple　Design Company: DC Alliance Architecture Design　Designers: Dong Yi, Cheng Jiujun, Zhao Wei, Jin Yuan　Site Area: 11,000m²　Building Area: 12,000m²
Completion Time: Jan. 2014　Location: Gongyuan Street, Qinhuai District, Nanjing, Jiangsu Province, China

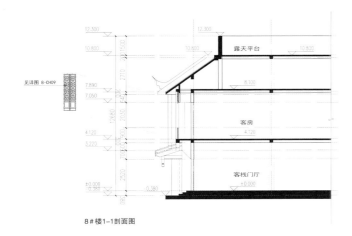

8#楼1-1剖面图

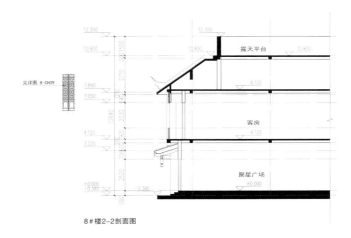

8#楼2-2剖面图

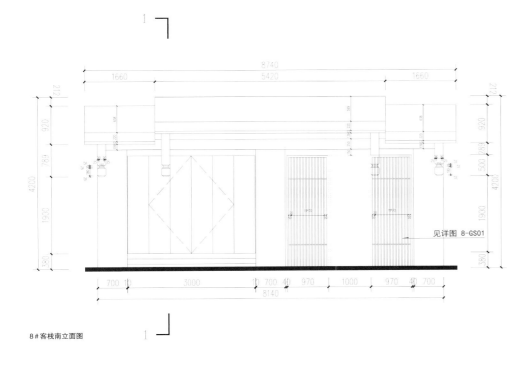

8#客栈南立面图

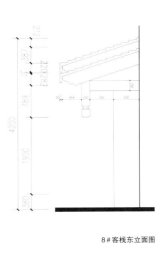

8#客栈东立面图

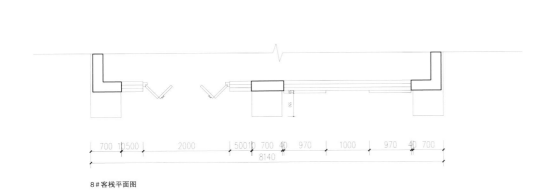

8#客栈平面图

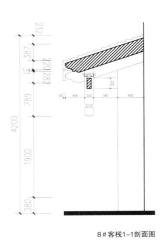

8#客栈1-1剖面图

南京夫子庙改造项目（一期）
The Confucian Temple Renovation Phase I, Nanjing

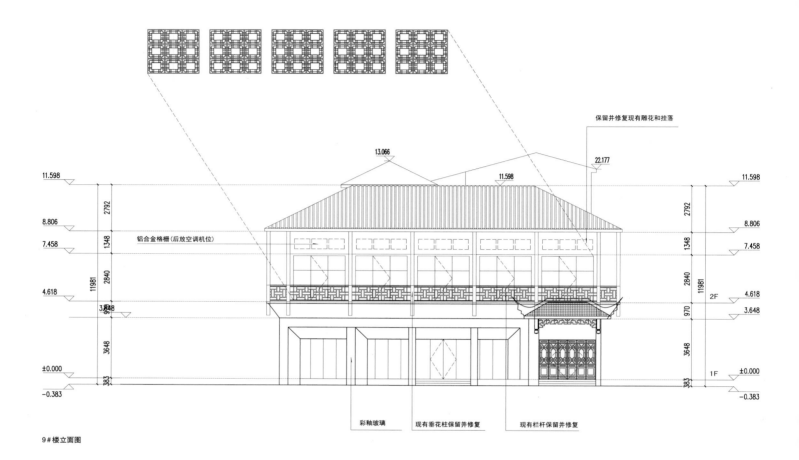

保留并修复现有雕花和挂落

铝合金格栅(后放空调机位)

彩釉玻璃   现有垂花柱保留并修复   现有栏杆保留并修复

9#楼立面图

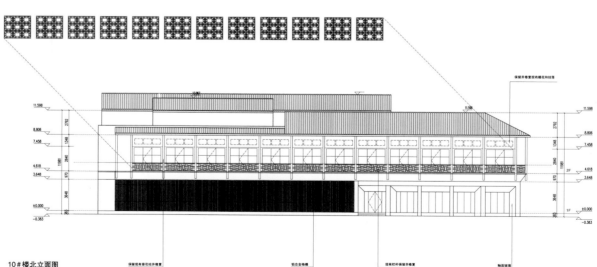

10#楼北立面图

业主：夫子庙管委会　设计单位：DC国际建筑设计事务所　设计师：董屹、程久军、赵维、金渊　用地面积：11 000平方米　建筑面积：12 000平方米　竣工时间：2014年1月　地点：南京市秦淮区贡院街
Client: The Administrative Committee of the Confucian Temple　Design Company: DC Alliance Architecture Design　Designers: Dong Yi, Cheng Jiujun, Zhao Wei, Jin Yuan　Site Area: 11,000m²　Building Area: 12,000m²
Completion Time: Jan. 2014　Location: Gongyuan Street, Qinhuai District, Nanjing, Jiangsu Province, China

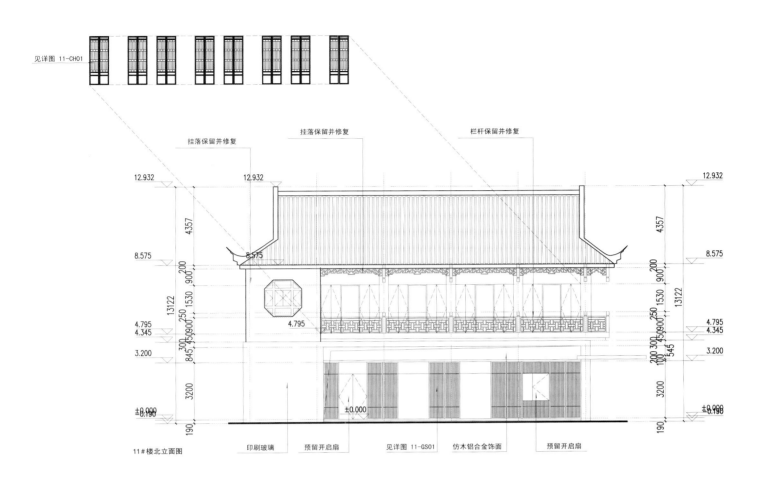

见详图 11-CH01

挂落保留并修复　　挂落保留并修复　　栏杆保留并修复

11#楼北立面图

印刷玻璃　　预留开启扇　　见详图 11-GS01　　仿木铝合金饰面　　预留开启扇

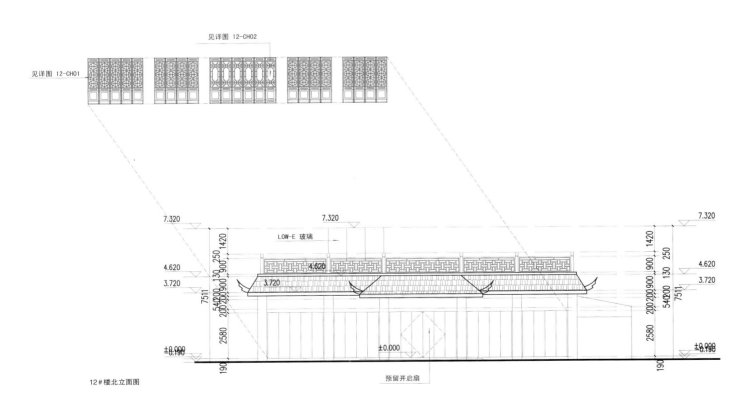

见详图 12-CH02

见详图 12-CH01

LOW-E 玻璃

12#楼北立面图

预留开启扇

重庆香霏古街
Xiangfei Ancient Street, Chongqing

重庆香霏古街
Xiangfei Ancient Street, Chongqing

地点：中国重庆市大足区　用地面积：30 000平方米　建筑面积：19 000平方米　业主：泽京集团　设计单位：DC国际建筑设计事务所
Location: Dazu District, Chongqing, China　Site Area: 30,000m²　Building Area: 19,000m²　Client: Zejing Group　Design Company: DC Alliance Architecture Design

重庆香霏古街
Xiangfei Ancient Street, Chongqing

生态绿化停车场

海棠森林公园

总平面图

地点：**中国重庆市大足区**　用地面积：**30 000平方米**　建筑面积：**19 000平方米**　业主：**泽京集团**　设计单位：**DC国际建筑设计事务所**
Location: Dazu District, Chongqing, China　Site Area: 30,000m²　Building Area: 19,000m²　Client: Zejing Group　Design Company: DC Alliance Architecture Design

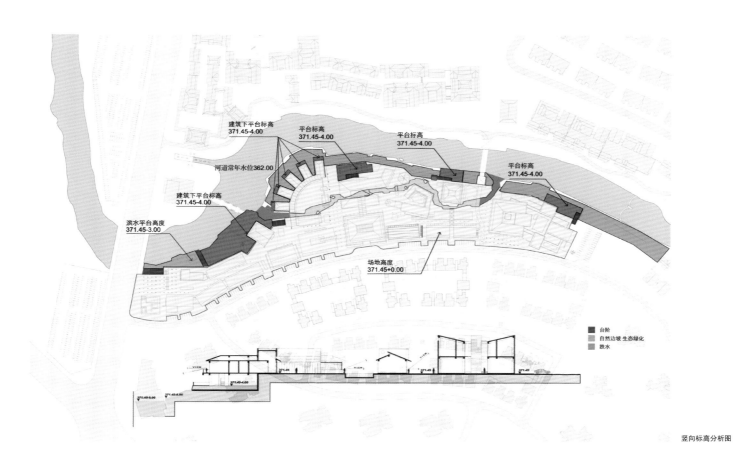

竖向标高分析图

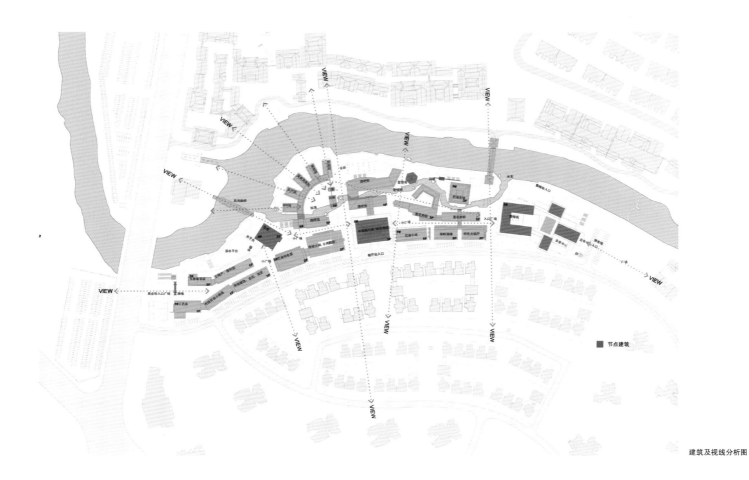

建筑及视线分析图

重庆香霏古街
Xiangfei Ancient Street, Chongqing

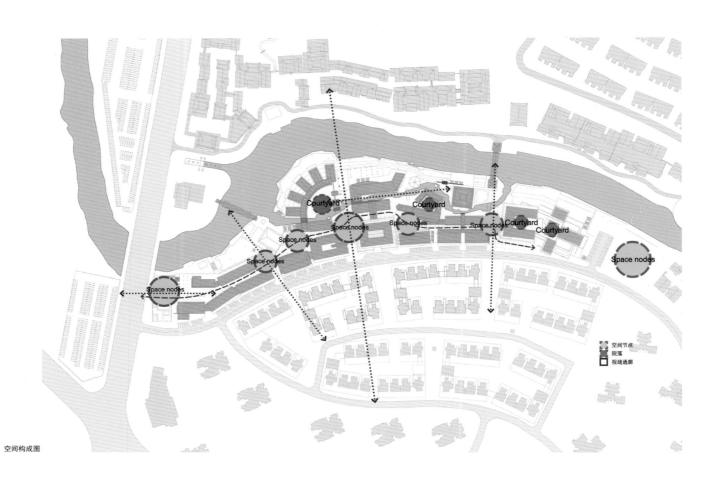

空间构成图

地点：**中国重庆市大足区**　用地面积：**30 000平方米**　建筑面积：**19 000平方米**　业主：**泽京集团**　设计单位：**DC国际建筑设计事务所**
Location: Dazu District, Chongqing, China　Site Area: 30,000m²　Building Area: 19,000m²　Client: Zejing Group　Design Company: DC Alliance Architecture Design

重庆香霏古街
Xiangfei Ancient Street, Chongqing

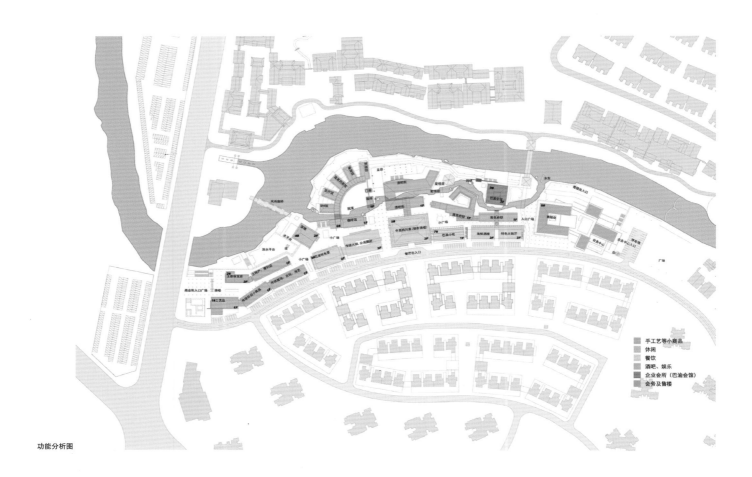

功能分析图

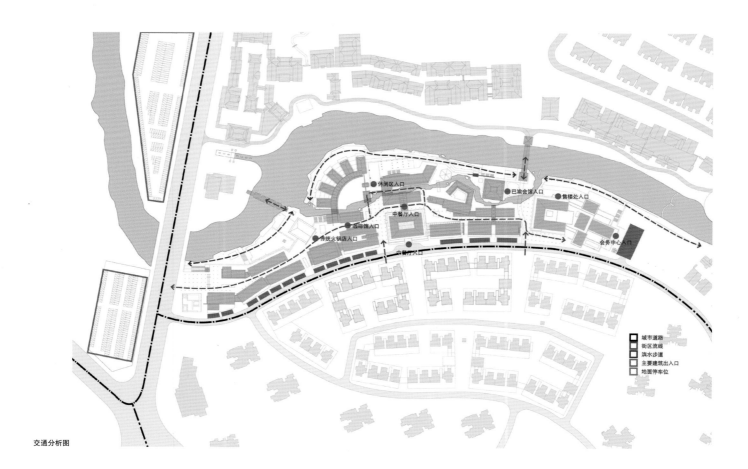

交通分析图

地点：中国重庆市大足区　用地面积：30 000平方米　建筑面积：19 000平方米　业主：泽京集团　设计单位：DC国际建筑设计事务所
Location: Dazu District, Chongqing, China　Site Area: 30,000m²　Building Area: 19,000m²　Client: Zejing Group　Design Company: DC Alliance Architecture Design

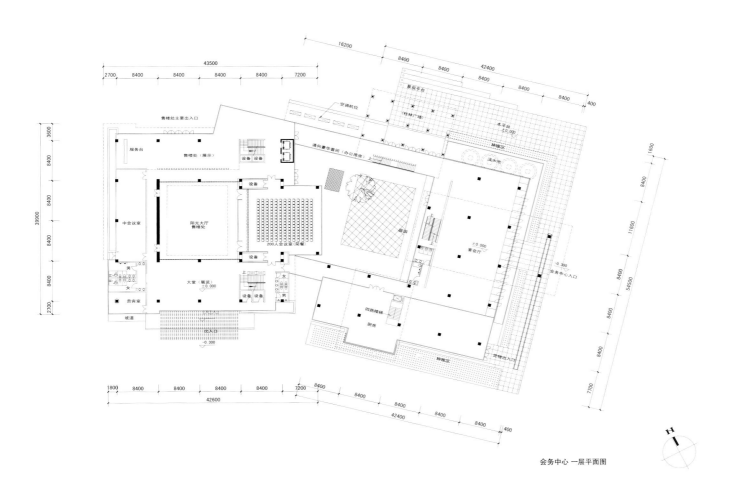

会务中心 一层平面图

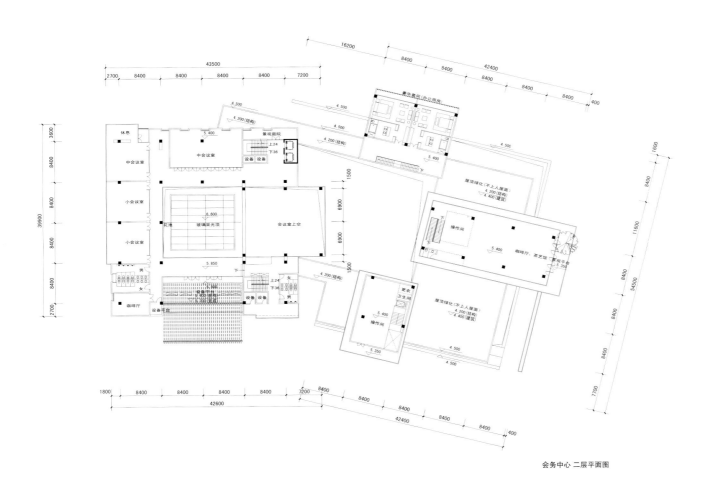

会务中心 二层平面图

重庆香霏古街
Xiangfei Ancient Street, Chongqing

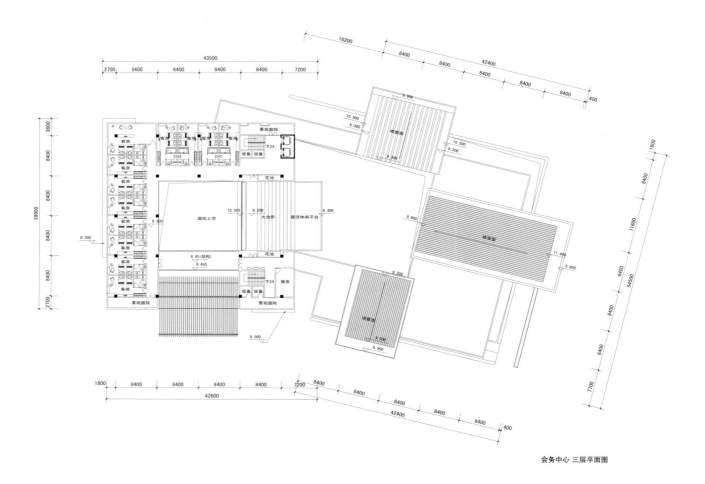

会务中心 三层平面图

地点：中国重庆市大足区　用地面积：30 000平方米　建筑面积：19 000平方米　业主：泽京集团　设计单位：DC国际建筑设计事务所
Location: Dazu District, Chongqing, China　Site Area: 30,000m²　Building Area: 19,000m²　Client: Zejing Group　Design Company: DC Alliance Architecture Design

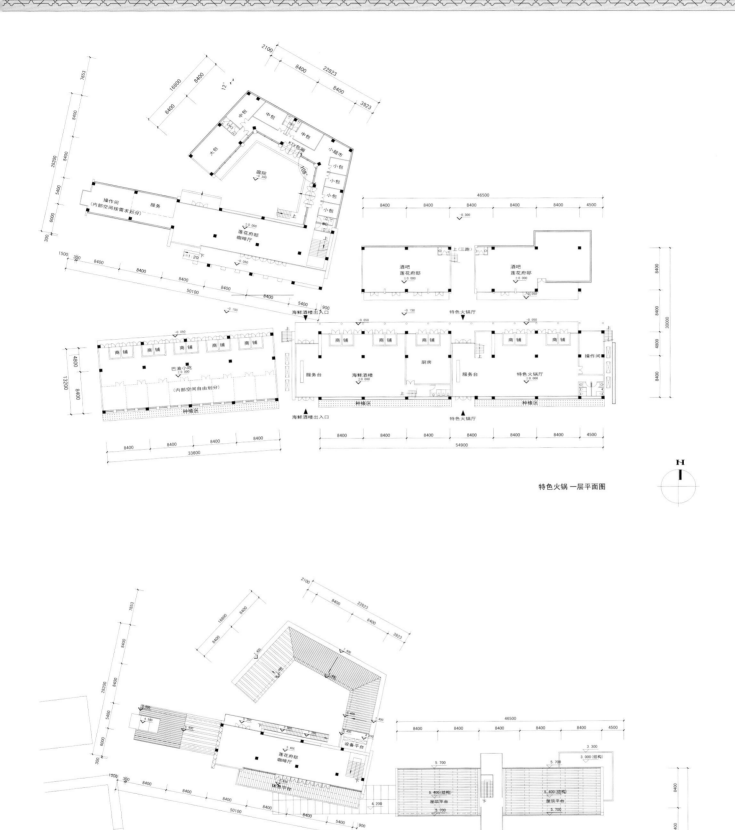

特色火锅 一层平面图

特色火锅 二层平面图

重庆香霏古街
Xiangfei Ancient Street, Chongqing

地点：中国重庆市大足区　用地面积：30 000平方米　建筑面积：19 000平方米　业主：泽京集团　设计单位：DC国际建筑设计事务所
Location: Dazu District, Chongqing, China　Site Area: 30,000m²　Building Area: 19,000m²　Client: Zejing Group　Design Company: DC Alliance Architecture Design

重庆香霏古街
Xiangfei Ancient Street, Chongqing

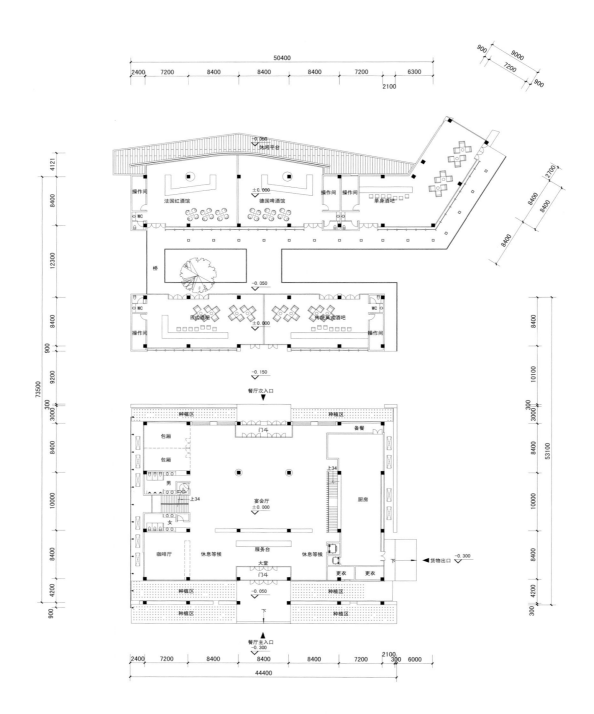

酒吧街 一层平面图

地点：中国重庆市大足区　用地面积：30 000平方米　建筑面积：19 000平方米　业主：泽京集团　设计单位：DC国际建筑设计事务所
Location: Dazu District, Chongqing, China　Site Area: 30,000m²　Building Area: 19,000m²　Client: Zejing Group　Design Company: DC Alliance Architecture Design

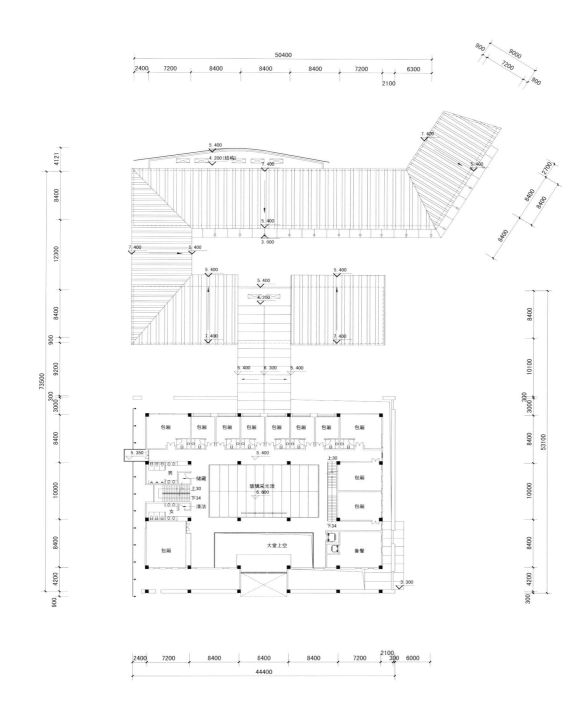

酒吧街 二层平面图

重庆香霏古街
Xiangfei Ancient Street, Chongqing

地点：中国重庆市大足区　用地面积：30 000平方米　建筑面积：19 000平方米　业主：泽京集团　设计单位：DC国际建筑设计事务所
Location: Dazu District, Chongqing, China　Site Area: 30,000m²　Building Area: 19,000m²　Client: Zejing Group　Design Company: DC Alliance Architecture Design

[解析商业街区与商业建筑设计]：商业街 & 社区商业　　[Design Analysis of Commercial Districts & Commercial Buildings]：Commercial Streets & Community Commerce

杭州万科良渚文化村"玉鸟流苏"
Vanke Jade Bird Ville, Liangzhu New Town Hangzhou

杭州万科良渚文化村"玉鸟流苏"
Vanke Jade Bird Ville, Liangzhu New Town Hangzhou

开发商：杭州万科　建筑设计：张雷联合建筑事务所（合作设计：南京大学建筑规划设计研究院有限公司）　设计师：张雷、肖育智、戚威、梅蕊蕊　建筑面积：8 000平方米　地点：中国浙江省杭州市良渚文化村　摄影：贾方
Developer: Hangzhou Vanke　Architecture Design: RZL Architects (in collaboration with ADINJU)　Designers: Zhang Lei, Xiao Yuzhi, Qi Wei, Mei Ruirui　Building Area: 8,000m²　Location: Liangzhu New Town, Hangzhou, Zhejiang Province, China　Photography: Jia Fang

杭州万科良渚文化村"玉鸟流苏"
Vanke Jade Bird Ville, Liangzhu New Town Hangzhou

总平面图

杭州万科良渚文化村"玉鸟流苏"
Vanke Jade Bird Ville, Liangzhu New Town Hangzhou

开发商：杭州万科　建筑设计：张雷联合建筑事务所（合作设计：南京大学建筑规划设计研究院有限公司）　设计师：张雷、肖育智、戚威、梅蕊蕊　建筑面积：8 000平方米　地点：中国浙江省杭州市良渚文化村　摄影：贾方
Developer: Hangzhou Vanke　Architecture Design: RZL Architects (in collaboration with ADINJU)　Designers: Zhang Lei, Xiao Yuzhi, Qi Wei, Mei Ruirui　Building Area: 8,000m²　Location: Liangzhu New Town, Hangzhou, Zhejiang Province, China　Photography: Jia Fang

概念分析图

方案的构想来自对基地上自然村落有机形态的分析。村落的基本构成单位是家庭，每个家庭以类似的宅院方式聚合。内与外的区分一方面表现在建筑南向与北向虚实的形态上，另一方面通过院落和天井来界定。类似的单元以自然生长的方式结合在一起，形成聚落，以个体内敛的姿态界定街巷，形成公共场所。

A、F地块的组成单元在类型学上有着相同的原型。自然村落中的普通家庭空间单元内与外辨证的关系在这里被传承下来，成为重构的出发点，而不是简单模仿白墙灰瓦一目了然的外形。密度的控制在总体关系中非常关键，不是担心被夸大，而是怕它们太小以至于不能限定街巷。

坡屋顶之所以在这里仍然被沿用，一是规划要求与周围建筑共同组成更大范围里相互和谐的整体，二是为内部空间提供未来多种使用的可能。庭院是设计的重点，它是建筑功能拓展的容器，也是内部活动的背景。通过今后对院墙不同围合材料与方式的选择，庭院可以转化为公共小广场，也可以成为完全封闭的辅助性后院，以满足今后可能改变的实用需要。

杭州万科良渚文化村"玉鸟流苏"
Vanke Jade Bird Ville, Liangzhu New Town Hangzhou

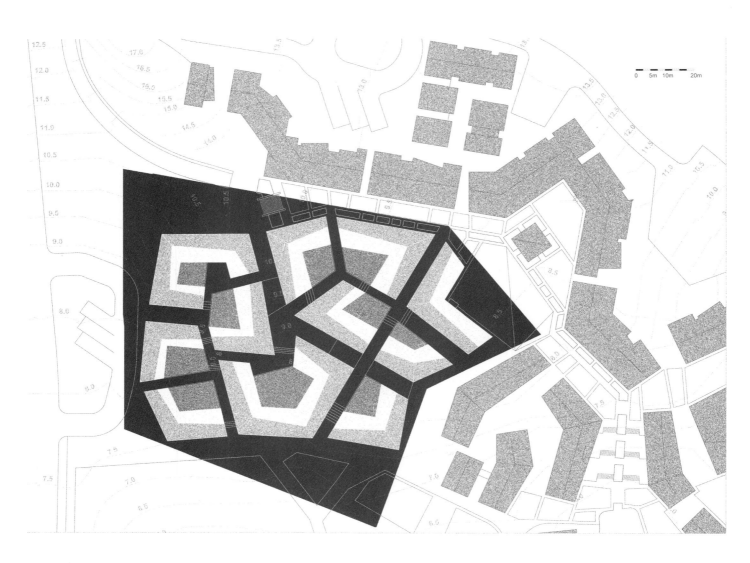

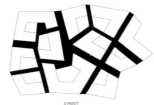

总平面分析图

The proposal is inspired from the analysis on the natural and organic village form on the site. The village is basically formed by families, and each family features similar houses with a courtyard. Interior and exterior are distinguished from each other by the buildings' south orientation and north orientation. Besides, courtyards often use walls to define interior and exterior. Similar families naturally grow together and form settlement.

The components of A and F lands have the same prototype in typology. Not only the typical appearance of white wall and gray tile is continued, but also the dialectical relationship between the interior and exterior of common family space in the natural village is inherited and becomes the start point of reconstruction. Density control is very essential in the general relationship. It's not a concern if it is magnified, but it is if it is too small to catalyze the defined street.

The reason why slope roof is used here is firstly to integrate into surrounding buildings as a harmonious entity, and secondly to create possibilities of multiple functions in the future for the interior space. Courtyard plays the most important role in the design. It is not only a container for building to extend its functions, but also a background for interior activities. Through different wall materials and different enclosure methods, courtyard can be transformed into a small public square, or completely enclosed backyard, meeting changeable function demands.

开发商: 杭州万科 建筑设计: 张雷联合建筑事务所(合作设计: 南京大学建筑规划设计研究院有限公司) 设计师: 张雷、肖育智、戚威、梅蕊蕊 建筑面积: 8 000平方米 地点: 中国浙江省杭州市良渚文化村 摄影: 贾方
Developer: Hangzhou Vanke Architecture Design: RZL Architects (in collaboration with ADINJU) Designers: Zhang Lei, Xiao Yuzhi, Qi Wei, Mei Ruirui Building Area: 8,000m² Location: Liangzhu New Town, Hangzhou, Zhejiang Province, China Photography: Jia Fang

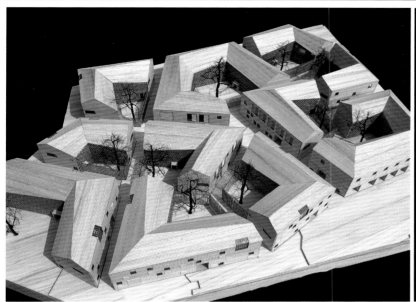

模型图

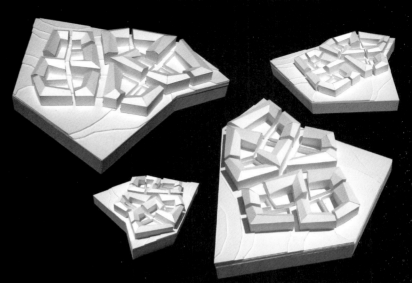

模型概念图

模型效果图

杭州万科良渚文化村"玉鸟流苏"
Vanke Jade Bird Ville, Liangzhu New Town Hangzhou

开发商: **杭州万科** 建筑设计: **张雷联合建筑事务所（合作设计: 南京大学建筑规划设计研究院有限公司）** 设计师: **张雷、肖育智、戚威、梅蕊蕊** 建筑面积: **8 000平方米** 地点: **中国浙江省杭州市良渚文化村** 摄影: **贾方**
Developer: Hangzhou Vanke   Architecture Design: RZL Architects (in collaboration with ADINJU)   Designers: Zhang Lei, Xiao Yuzhi, Qi Wei, Mei Ruirui   Building Area: 8,000m²   Location: Liangzhu New Town, Hangzhou, Zhejiang Province, China   Photography: Jia Fang

开发商: **杭州万科** 建筑设计: **张雷联合建筑事务所（合作设计: 南京大学建筑规划设计研究院有限公司）** 设计师: **张雷、肖育智、戚威、梅蕊蕊** 建筑面积: **8 000平方米** 地点: **中国浙江省杭州市良渚文化村** 摄影: **贾方**
Developer: Hangzhou Vanke   Architecture Design: RZL Architects (in collaboration with ADINJU)   Designers: Zhang Lei, Xiao Yuzhi, Qi Wei, Mei Ruirui   Building Area: 8,000m²   Location: Liangzhu New Town, Hangzhou, Zhejiang Province, China   Photography: Jia Fang

[解析商业街区与商业建筑设计]：商业街＆社区商业　[Design Analysis of Commercial Districts & Commercial Buildings] : Commercial Streets & Community Commerce

杭州万科良渚文化村 "玉鸟流苏"
Vanke Jade Bird Ville, Liangzhu New Town Hangzhou

地点：中国四川省成都市　用地面积：305 674.54平方米　建筑面积：1 890 000平方米　业主：成都龙湖锦鸿置业有限公司　建筑设计：承构建筑　主创设计：唐聃　参与设计：黄珏磊
Location: Chengdu, Sichuan Province, China　Site Area: 305,674.54m²　Building Area: 1,890,000m²　Client: Chengdu Longfor Jinrong Properties Co., Ltd.　Architecture Design: Made Make Architecture　Chief Designer: Tang Dan
Participated Designer: Huang Juelei

N

NW NE

W E

42

SW SE

S

成都龙湖时代天街
Longfor Times Paradise Walk, Chengdu

设计师在用地的四个拐角部位设置了数码港、餐饮楼、百货及天街四个大型商业，大型商业之间通过情景商业穿插衔接。

集中商业在空间设计中采用了天街的形式与情景商业连接，形成商业水平环线，增加了空间的丰富性和购物的趣味性。

塔楼分布在集中商业及情景商业上部，主要位于地块西区，东区避开影院及冰场对上部塔楼的影响，使整个空间疏密有致，塔楼在设计中更加考虑空间的增值与扩展，通过远近搭配、形体转折等，形成近大远小、阴暗交错的关系，丰富城市天际线。

对于多个面向城市节点的出入口，设计以营造一个该区域的标志性建筑为出发点，加强整个项目的可识别性。项目在设计上力求创新，充分体现天街的概念，以崭新的设计理念打造城西区域新中心，使其成为成都除宽窄巷子、锦里外的另一张城市名片。

地点：中国四川省成都市　用地面积：305 674.54平方米　建筑面积：1 890 000平方米　业主：成都龙湖锦鸿置业有限公司　建筑设计：承构建筑　主创设计：唐聃　参与设计：黄珏磊
Location: Chengdu, Sichuan Province, China　Site Area: 305,674.54m²　Building Area: 1,890,000m²　Client: Chengdu Longfor Jinrong Properties Co., Ltd.　Architecture Design: Made Make Architecture　Chief Designer: Tang Dan
Participated Designer: Huang Juelei

成都龙湖时代天街
Longfor Times Paradise Walk, Chengdu

地点：中国四川省成都市　用地面积：305 674.54平方米　建筑面积：1 890 000平方米　业主：成都龙湖锦鸿置业有限公司　建筑设计：承构建筑　主创设计：唐聃　参与设计：黄珏磊
Location: Chengdu, Sichuan Province, China　Site Area: 305,674.54m²　Building Area: 1,890,000m²　Client: Chengdu Longfor Jinrong Properties Co., Ltd.　Architecture Design: Made Make Architecture　Chief Designer: Tang Dan
Participated Designer: Huang Juelei

成都龙湖时代天街
Longfor Times Paradise Walk, Chengdu

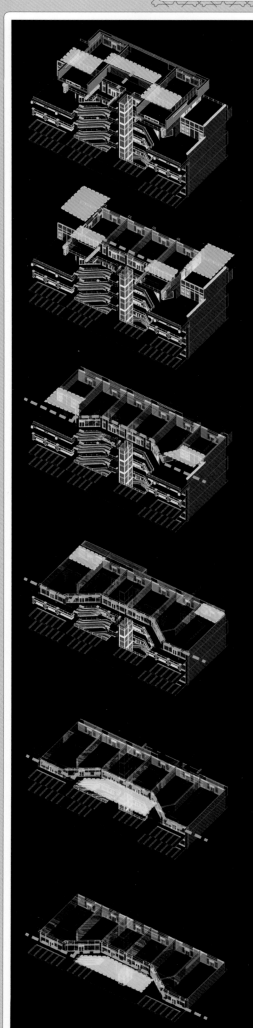

5、6F

人流最少，商业价值也最小，设置空中四合院，既增加了赠送面积，又丰富了空间趣味和立面造型，最大程度地提高空间价值。

4F

人流量进一步减少，商业价值再次降低，以赠送更多空中花园的形式增加商铺价值。

3F

仅有临近广场或内街口部的局部设置1F直通3F的自动扶梯，并补充点式的垂直电梯，使跨街的距离应尽量做短。

同时，三层的人流量减少，使商业价值降低，以赠送空中花园的形式增加商铺的价值。

2F

端头商铺面积较大，而位置欠佳，以赠送一定面积的空中花园来提高价值。

1F

是人流最为集中的楼层，关键是如何将其引入-1F。通向-1F的楼梯起点设置在核心位置，楼梯周边有开放的硬质铺地，便于集散。两个下层庭院尺度上适度平衡，在交通布置上尽量减少对后方商业店面的遮挡。

临街商铺商业价值最高，尽量做到最大密度。

B1F

交通汇集，相对集中，保证一定的商业人流，设计在考虑店面收益均衡的前提下使2F动线更为便捷。

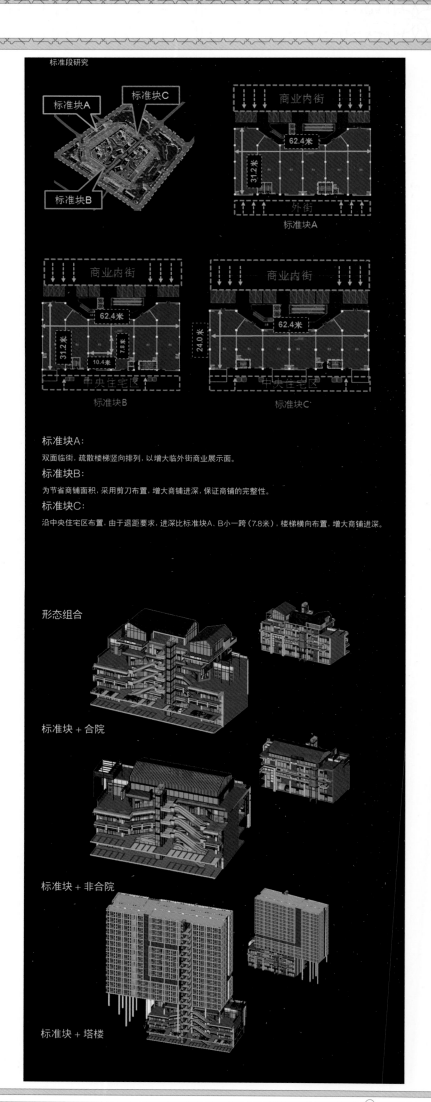

标准段研究

标准块A：

双面临街，疏散楼梯竖向排列，以增大临外街商业展示面。

标准块B：

为节省商铺面积，采用剪刀布置，增大商铺进深，保证商铺的完整性。

标准块C：

沿中央住宅区布置，由于退距要求，进深比标准块A、B小一跨（7.8米），楼梯横向布置，增大商铺进深。

形态组合

标准块 + 合院

标准块 + 非合院

标准块 + 塔楼

地点：中国四川省成都市　用地面积：305 674.54平方米　建筑面积：1 890 000平方米　业主：成都龙湖锦鸿置业有限公司　建筑设计：承构建筑　主创设计：唐聃　参与设计：黄珏磊

Location: Chengdu, Sichuan Province, China　Site Area: 305,674.54m²　Building Area: 1,890,000m²　Client: Chengdu Longfor Jinrong Properties Co., Ltd.　Architecture Design: Made Make Architecture　Chief Designer: Tang Dan

Participated Designer: Huang Juelei

26#塔楼衍变

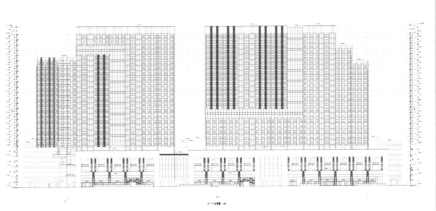

部分窗下墙和竖向空调百叶变为橙色 →

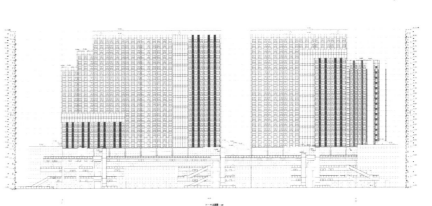

成都龙湖时代天街
Longfor Times Paradise Walk, Chengdu

地点：中国四川省成都市 用地面积：305 674.54平方米 建筑面积：1 890 000平方米 业主：成都龙湖锦鸿置业有限公司 建筑设计：承构建筑 主创设计：唐聃 参与设计：黄珏磊
Location: Chengdu, Sichuan Province, China   Site Area: 305,674.54m² Building Area: 1,890,000m²   Client: Chengdu Longfor Jinrong Properties Co., Ltd.   Architecture Design: Made Make Architecture   Chief Designer: Tang Dan
Participated Designer: Huang Juelei

Overview

The project is the largest one-stop shopping center in current Asia, and is ranked in the top three of the world, and it is a commercial project with the greatest investment value across Chengdu. It is a super commercial complex that is beyond your imagination, with a site area of 305,674.54m² and a building area of 1,890,000m².

The project is situated in Gaoxinxi District, the only urban core zone which gathers central residential area, central education area and national bounded area in Chengdu's main urban area. It captures a large commercial land that is extremely scarce in Gaoxinxi District, meets the various consumption demands from west Chengdu to Pixian County, and enjoys approximate two million consumers in total, which come from the large residential communities in west Chengdu and Pixian County, and Chengdu's largest university town, and Chengdu's only comprehensive bounded area. The project will soon become a trans-regional commercial center which monopolizes west Chengdu, radiates across the whole Chengdu, and even influences southwest China.

The project features an all-round three-dimensional traffic network, formed by subway line 2, Chengdu-Dujiangyan Express Railway, Chengdu-Dujiangyan Expressway, IT Avenue and the highly developed road network of Gaoxinxi District, seamlessly connected with Chengdu's main urban area, Pixian County and Dujiangyan.

The project houses super large anchor department store, Longfor Paradise Walk shopping center, MOCO furniture area, food world, etc. becoming an integrated one-stop consumption center. It is composed of scenic commercial street stores, hardbound SOHO apartment, creative loft and hardbound residence, most of which are commercial. What's more, seven urban functions including leisure & shopping, administration & office, F&B & entertainment, star hotel, creative industry, urban plaza, and culture & art are included in the project.

Planning Concept

The project has a plane circulation that features a combination of ring-shape shopping path and sky walk. It also has a three-dimension circulation: two rings—scenic commercial street and golden inner ring, gathering customer flow and enhancing commercial atmosphere; three traverses—dozens of transverse sky bridges that cross above the street on the second, third and fourth floors of the scenic commercial street, allowing customer flow to move unimpededly; four verticals—sunken square, escalator, cross-floor escalator and vertical panorama elevator, distributing customer flow vertically.

The project sets open space on the clear street mainline. From road to space, from line to plane, multi-layer experiences are created. The scenic commercial street adopts standard length, and it not only ensures the integrity of commercial facade, but also adds interestingness to the inner space. The street commerce continuously runs for more than 2,000m long, and important commercial nodes are set at every 500m, while small commercial nodes are set at every 100m.

At the four corners of the site, cyberport, F&B building, department store and sky walk are installed, and these four large businesses are connected by scenic commerce.

The concentrated commerce is framed within sky walks, thus to connect scenic commerce, bringing richness to the space and interestingness to shopping.

The towers, mainly located in the west of the site, sit above concentrated commerce and scenic commerce. In the east of the site, towers are reduced in quantity, so as to avoid the influence imposed upon them by cinema and skating rink. As a result, a space sense with proper density is created. Besides, the tower design pays great attention to the space appreciation and expansion. As for the several entrances & exits that open to the city, the design concept starts from creating a landmark building for the region, so as to strengthen the identity of the project. What's more, the design strives to be innovative and fully embodies the concept of paradise walk, aiming to become a new center of west Chengdu, and another city card apart from Kuan Zhai Alley and Jinliwai.

成都龙湖时代天街
Longfor Times Paradise Walk, Chengdu

地点：中国四川省成都市 用地面积：305 674.54平方米 建筑面积：1 890 000平方米 业主：成都龙湖锦鸿置业有限公司 建筑设计：承构建筑 主创设计：唐聃 参与设计：黄珏磊
Location: Chengdu, Sichuan Province, China Site Area: 305,674.54m² Building Area: 1,890,000m² Client: Chengdu Longfor Jinrong Properties Co., Ltd. Architecture Design: Made Make Architecture Chief Designer: Tang Dan
Participated Designer: Huang Juelei

成都龙湖时代天街
Longfor Times Paradise Walk, Chengdu

成都龙湖时代天街
Longfor Times Paradise Walk, Chengdu

地点：中国四川省成都市　用地面积：305 674.54平方米　建筑面积：1 890 000平方米　业主：成都龙湖锦鸿置业有限公司　建筑设计：承构建筑　主创设计：唐聃　参与设计：黄珏磊
Location: Chengdu, Sichuan Province, China　Site Area: 305,674.54m²　Building Area: 1,890,000m²　Client: Chengdu Longfor Jinrong Properties Co., Ltd.　Architecture Design: Made Make Architecture　Chief Designer: Tang Dan
Participated Designer: Huang Juelei

成都龙湖时代天街
Longfor Times Paradise Walk, Chengdu

地点：中国四川省成都市　用地面积：305 674.54平方米　建筑面积：1 890 000平方米　业主：成都龙湖锦鸿置业有限公司　建筑设计：承构建筑　主创设计：唐聃　参与设计：黄珏磊
Location: Chengdu, Sichuan Province, China　Site Area: 305,674.54m²　Building Area: 1,890,000m²　Client: Chengdu Longfor Jinrong Properties Co., Ltd.　Architecture Design: Made Make Architecture　Chief Designer: Tang Dan
Participated Designer: Huang Juelei

成都龙湖时代天街
Longfor Times Paradise Walk, Chengdu

地点：中国四川省成都市　用地面积：305 674.54平方米　建筑面积：1 890 000平方米　业主：成都龙湖锦鸿置业有限公司　建筑设计：承构建筑　主创设计：唐骋　参与设计：黄珏磊
Location: Chengdu, Sichuan Province, China　Site Area: 305,674.54m²　Building Area: 1,890,000m²　Client: Chengdu Longfor Jinrong Properties Co., Ltd.　Architecture Design: Made Make Architecture　Chief Designer: Tang Dan
Participated Designer: Huang Juelei

扬州虹桥坊
Rainbow Square, Yangzhou

开发商：扬州瘦西湖旅游发展集团有限公司　建筑设计：上海都设建筑设计有限公司（配合单位：江苏时代建筑设计院有限公司）
Developer: Yangzhou Slender West Lake Tourism & Development Group　Architecture Design: Shanghai Dushe Architecture Design
Location: 18 Dahongqiao Road, Guangling District, Yangzhou, Jiangsu Province, China

扬州虹桥坊
Rainbow Square, Yangzhou
开发商：扬州瘦西湖旅游发展集团有限公司　建筑设计：上海都设建筑设计有限公司（配合单位：江苏时代建筑设计院有限公司）
Location: 18 Dahongqiao Road, Guangling District, Yangzhou, Jiangsu Province, China

景观设计：上海都设建筑设计有限公司（配合单位：扬州园林设计院有限公司）　占地面积：48 000平方米　建筑面积：地上25 000平方米　容积率：0.52　绿化率：32%　摄影：凌克戈　地点：江苏省扬州市广陵区大虹桥路18号
Co., Ltd.　Landscape Design: Shanghai Dushe Architecture Design Co., Ltd.　Site Area: 48,000m²　Building Area: 25,000m² (above ground)　Plot Ratio: 0.52　Greening Rate: 32%　Photography: Ling Kege

扬州虹桥坊
Rainbow Square, Yangzhou

开发商：扬州瘦西湖旅游发展集团有限公司　建筑设计：上海都设建筑设计有限公司（配合单位：江苏时代建筑设计院有限公司）
Developer: Yangzhou Slender West Lake Tourism & Development Group　Architecture Design: Shanghai Dushe Architecture Design
Location: 18 Dahongqiao Road, Guangling District, Yangzhou, Jiangsu Province, China

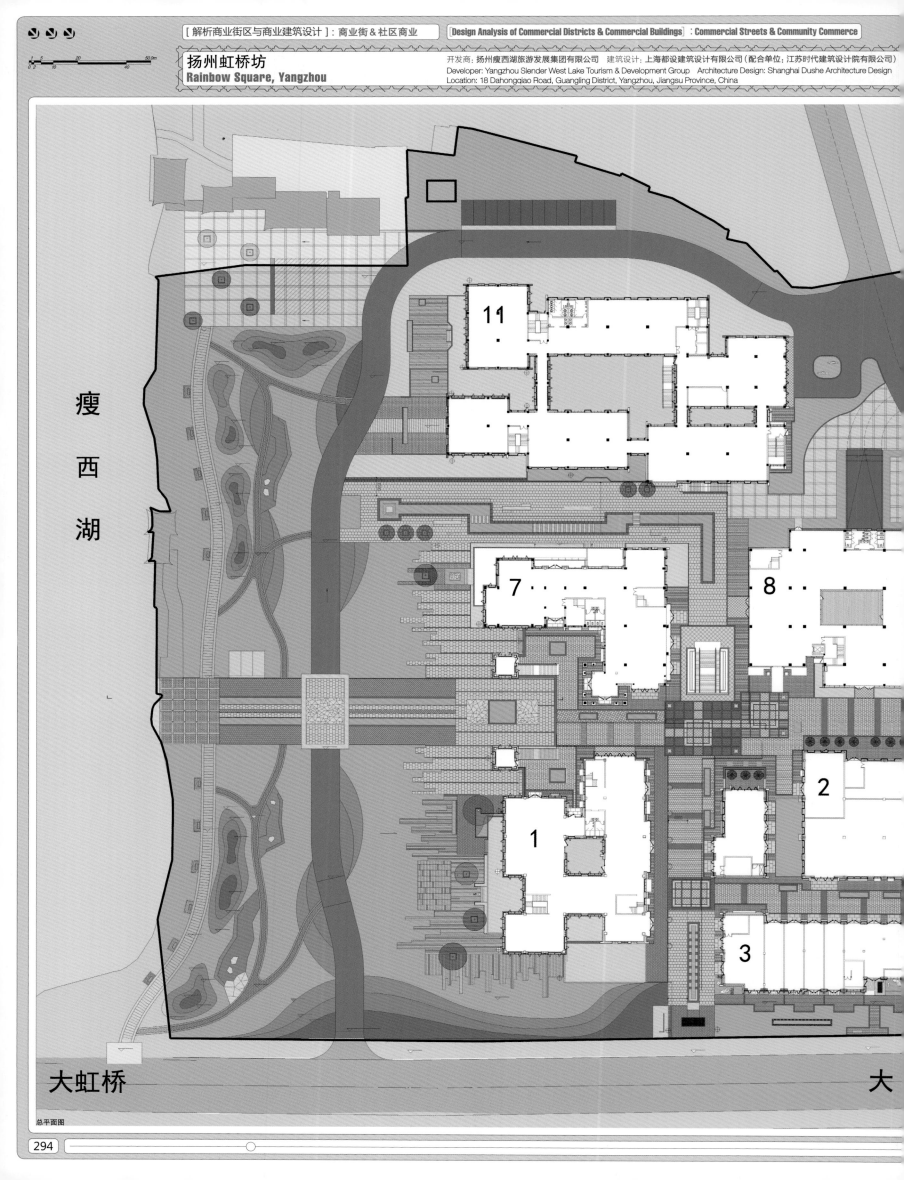

瘦西湖

大虹桥

大

总平面图

景观设计：上海都设建筑设计有限公司（配合单位：扬州园林设计院有限公司）　占地面积：48 000平方米　建筑面积：地上25 000平方米　容积率：0.52　绿化率：32%　摄影：凌克戈　地点：江苏省扬州市广陵区大虹桥路18号
Co., Ltd.　Landscape Design: Shanghai Dushe Architecture Design Co., Ltd.　Site Area: 48,000m²　Building Area: 25,000m² (above ground)　Plot Ratio: 0.52　Greening Rate: 32%　Photography: Ling Kege

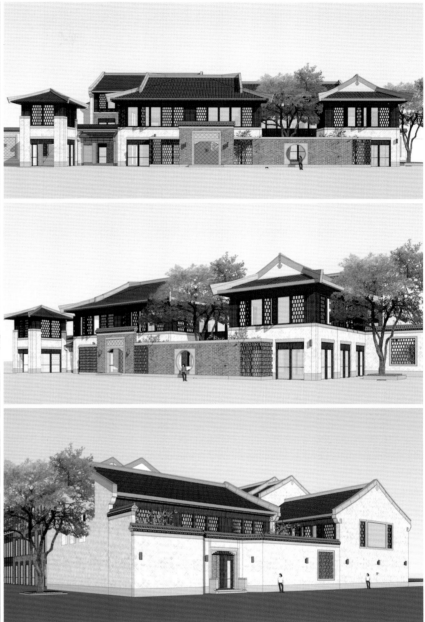

扬州虹桥坊
Rainbow Square, Yangzhou

开发商：扬州瘦西湖旅游发展集团有限公司　建筑设计：上海都设建筑设计有限公司（配合单位：江苏时代建筑设计院有限公司）
Developer: Yangzhou Slender West Lake Tourism & Development Group　Architecture Design: Shanghai Dushe Architecture Design
Location: 18 Dahongqiao Road, Guangling District, Yangzhou, Jiangsu Province, China

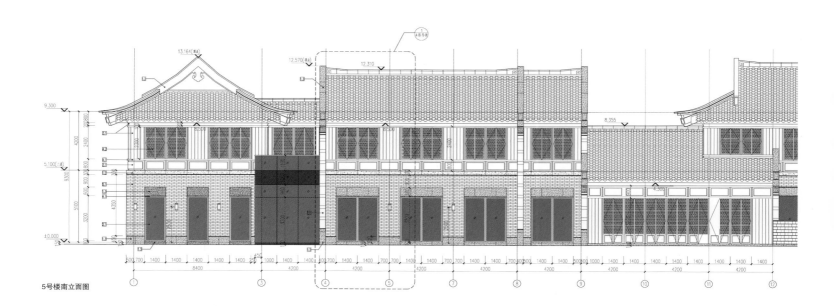

5号楼南立面图

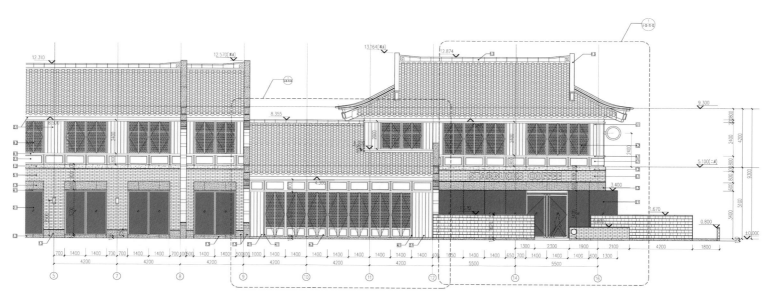

5号楼南立面图

景观设计: 上海都设建筑设计有限公司(配合单位: 扬州园林设计院有限公司)　占地面积: 48 000平方米　建筑面积: 地上25 000平方米　容积率: 0.52　绿化率: 32%　摄影: 凌克戈　地点: 江苏省扬州市广陵区大虹桥路18号
Co., Ltd.　Landscape Design: Shanghai Dushe Architecture Design Co., Ltd.　Site Area: 48,000m²　Building Area: 25,000m² (above ground)　Plot Ratio: 0.52　Greening Rate: 32%　Photography: Ling Kege

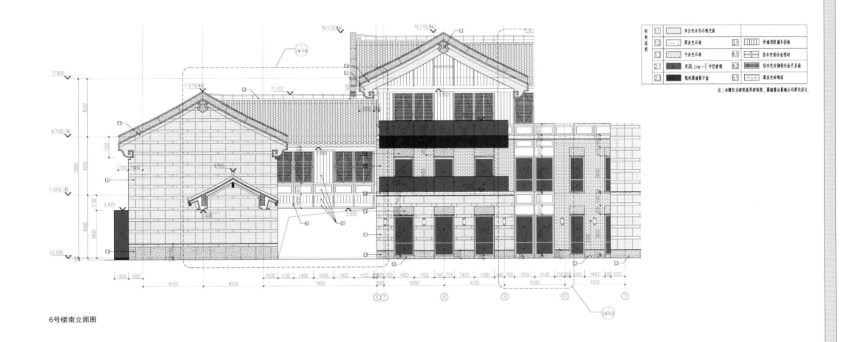

6号楼南立面图

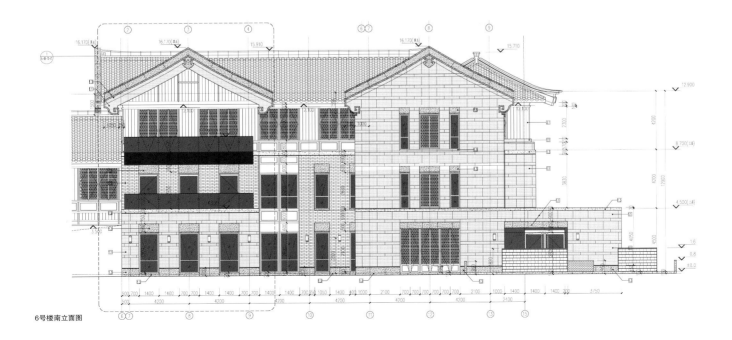

6号楼南立面图

扬州虹桥坊
Rainbow Square, Yangzhou

开发商：扬州瘦西湖旅游发展集团有限公司　建筑设计：上海都设建筑设计有限公司（配合单位：江苏时代建筑设计院有限公司）
Developer: Yangzhou Slender West Lake Tourism & Development Group　Architecture Design: Shanghai Dushe Architecture Design
Location: 18 Dahongqiao Road, Guangling District, Yangzhou, Jiangsu Province, China

景观设计：上海都设建筑设计有限公司（配合单位：扬州园林设计院有限公司）　占地面积：48 000平方米　建筑面积：地上25 000平方米　容积率：0.52　绿化率：32%　摄影：凌克戈　地点：江苏省扬州市广陵区大虹桥路18号
Co., Ltd.　Landscape Design: Shanghai Dushe Architecture Design Co., Ltd.　Site Area: 48,000m²　Building Area: 25,000m² (above ground)　Plot Ratio: 0.52　Greening Rate: 32%　Photography: Ling Kege

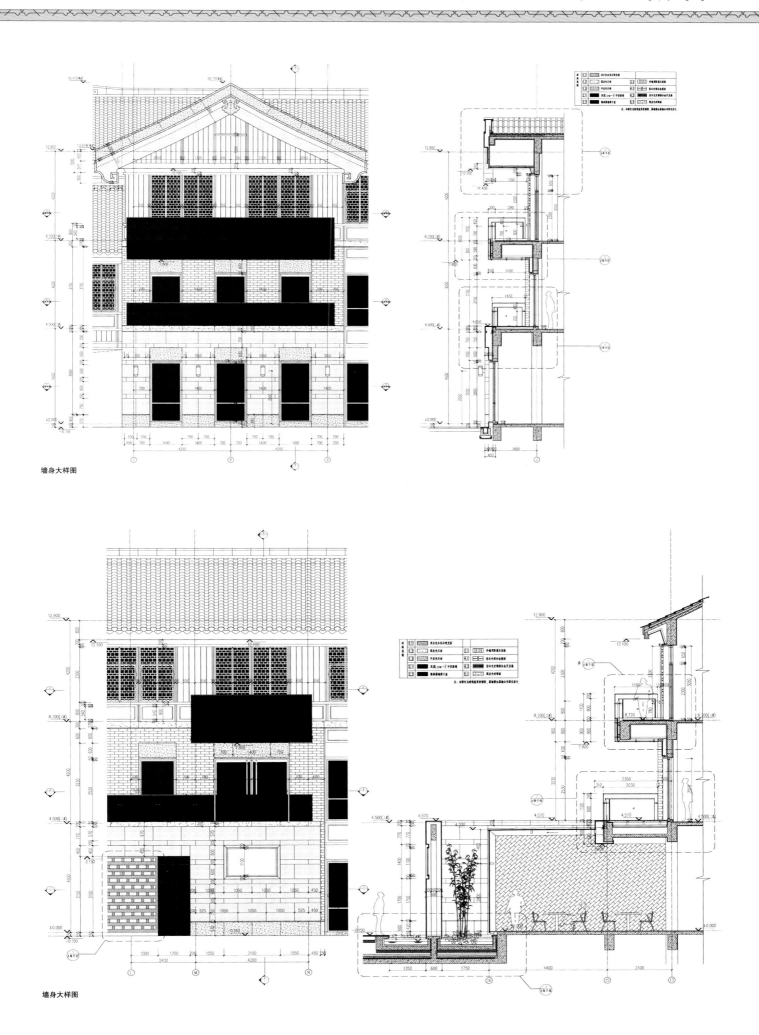

墙身大样图

墙身大样图

扬州虹桥坊
Rainbow Square, Yangzhou

开发商：扬州瘦西湖旅游发展集团有限公司   建筑设计：上海都设建筑设计有限公司（配合单位：江苏时代建筑设计院有限公司）
Developer: Yangzhou Slender West Lake Tourism & Development Group   Architecture Design: Shanghai Dushe Architecture Design
Location: 18 Dahongqiao Road, Guangling District, Yangzhou, Jiangsu Province, China

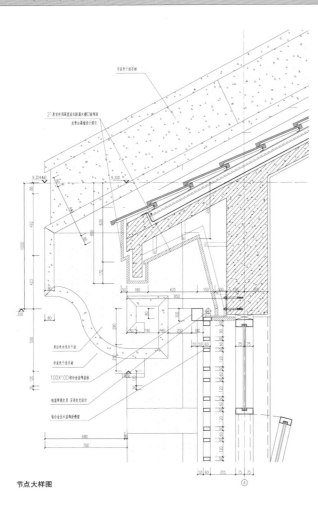

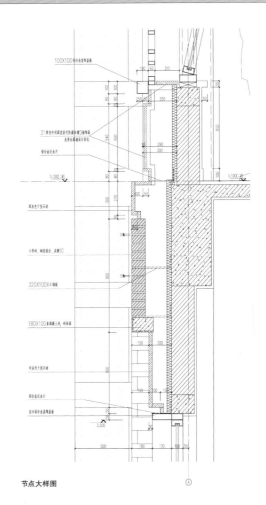

节点大样图

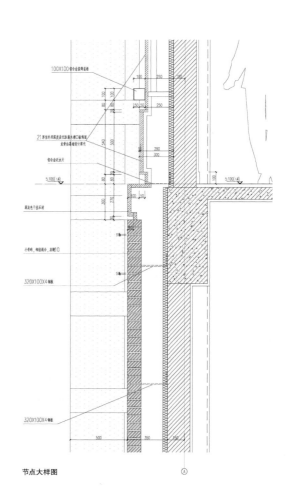

节点大样图

景观设计：上海都设建筑设计有限公司（配合单位：扬州园林设计院有限公司）　占地面积：48 000平方米　建筑面积：地上25 000平方米　容积率：0.52　绿化率：32%　摄影：凌克戈　地点：江苏省扬州市广陵区大虹桥路18号
Co., Ltd.　Landscape Design: Shanghai Dushe Architecture Design Co., Ltd.　Site Area: 48,000m²　Building Area: 25,000m² (above ground)　Plot Ratio: 0.52　Greening Rate: 32%　Photography: Ling Kege

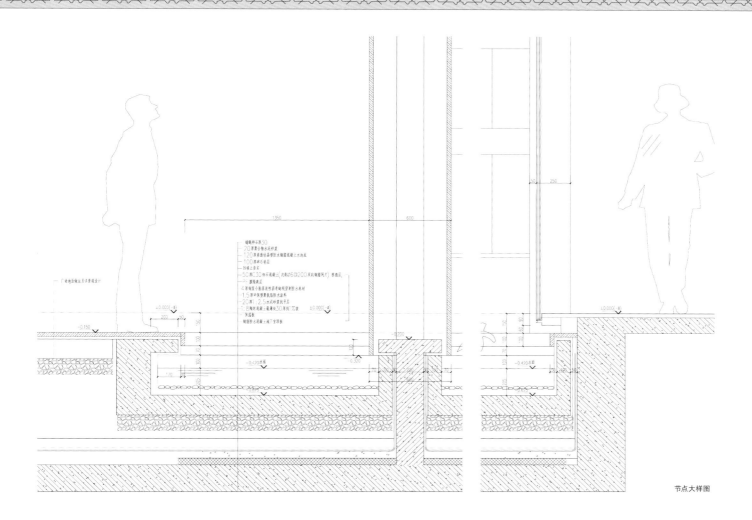

节点大样图

景观设计：上海都设建筑设计有限公司（配合单位：扬州园林设计院有限公司）　占地面积：48 000平方米　建筑面积：地上25 000平方米　容积率：0.52　绿化率：32%　摄影：凌克戈　地点：江苏省扬州市广陵区大虹桥路18号
Co., Ltd.　Landscape Design: Shanghai Dushe Architecture Design Co., Ltd.　Site Area: 48,000m²　Building Area: 25,000m² (above ground)　Plot Ratio: 0.52　Greening Rate: 32%　Photography: Ling Kege

扬州虹桥坊
Rainbow Square, Yangzhou

开发商：扬州瘦西湖旅游发展集团有限公司　建筑设计：上海都设建筑设计有限公司 (配合单位：江苏时代建筑设计院有限公司)
Developer: Yangzhou Slender West Lake Tourism & Development Group　Architecture Design: Shanghai Dushe Architecture Design
Location: 18 Dahongqiao Road, Guangling District, Yangzhou, Jiangsu Province, China

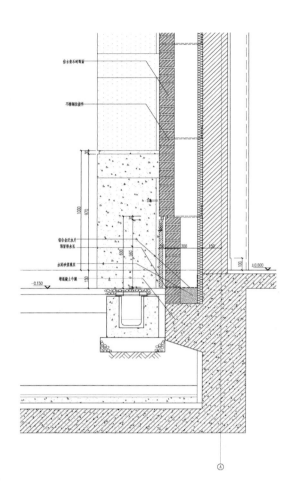

节点大样图

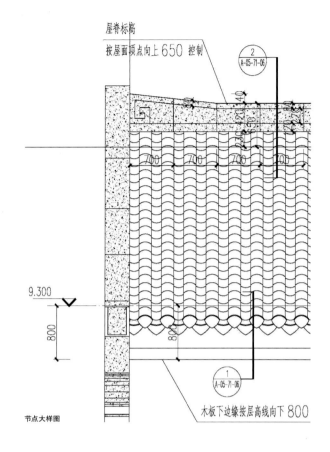

节点大样图

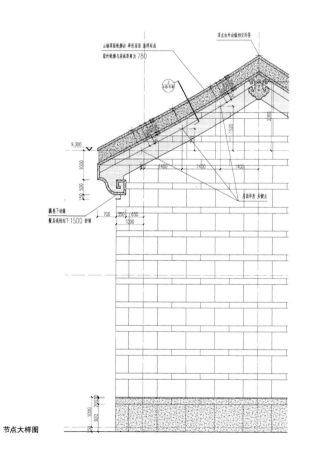

节点大样图

景观设计：上海都设建筑设计有限公司（配合单位：扬州园林设计院有限公司）　占地面积：48 000平方米　建筑面积：地上25 000平方米　容积率：0.52　绿化率：32%　摄影：凌克戈　地点：江苏省扬州市广陵区大虹桥路18号
Co., Ltd.　Landscape Design: Shanghai Dushe Architecture Design Co., Ltd.　Site Area: 48,000m²　Building Area: 25,000m² (above ground)　Plot Ratio: 0.52　Greening Rate: 32%　Photography: Ling Kege

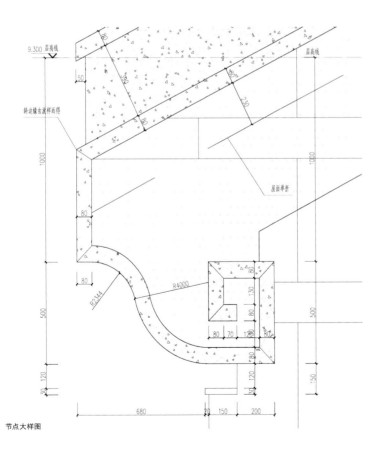

节点大样图

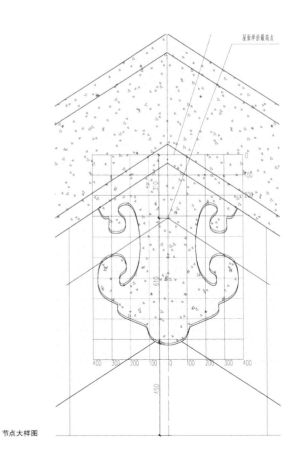

节点大样图

扬州虹桥坊
Rainbow Square, Yangzhou

开发商：扬州瘦西湖旅游发展集团有限公司   建筑设计：上海都设建筑设计有限公司（配合单位：江苏时代建筑设计院有限公司）
Developer: Yangzhou Slender West Lake Tourism & Development Group   Architecture Design: Shanghai Dushe Architecture Design
Location: 18 Dahongqiao Road, Guangling District, Yangzhou, Jiangsu Province, China

景观设计：上海都设建筑设计有限公司（配合单位：扬州园林设计院有限公司）　占地面积：48 000平方米　建筑面积：地上25 000平方米　容积率：0.52　绿化率：32%　摄影：凌克戈　地点：江苏省扬州市广陵区大虹桥路18号
Co., Ltd.　Landscape Design: Shanghai Dushe Architecture Design Co., Ltd.　Site Area: 48,000m²　Building Area: 25,000m² (above ground)　Plot Ratio: 0.52　Greening Rate: 32%　Photography: Ling Kege

宁波莲桥街
Lianqiao Street, Ningbo

占地面积：64 000平方米　建筑面积：118 184.8平方米　地点：中国浙江省宁波市
Site Area: 64,000m²　Building Area: 118,184.8m²　Location: Ningbo, Zhejiang Province, China

宁波莲桥街
Lianqiao Street, Ningbo

总平面图

占地面积：64 000平方米　建筑面积：118 184.8平方米　地点：中国浙江省宁波市
Site Area: 64,000m²　Building Area: 118,184.8m²　Location: Ningbo, Zhejiang Province, China

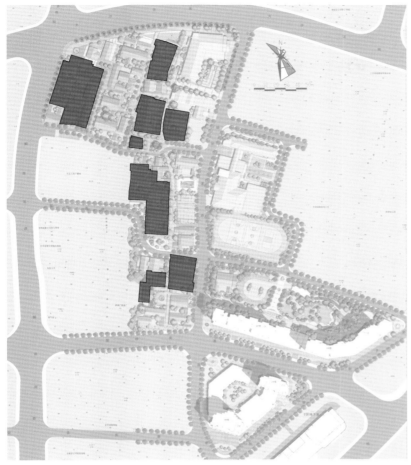

保留院落分析图

保护院落

现状肌理

恢复原有院落分析图

项目规划功能区包括传统街巷风情街、精品商业、会所、学校、高端公寓等，是一个典型的历史街区综合开发与更新的城市案例。项目旨在打造宁波中心城高端综合性的物业，并成为宁波市高端住宅发展中重要的里程碑，并成为旧城改造与历史风貌保护的新典范。

规划理念

保留一切可保留的院落；尊重原有肌理，迁建一些重要院落；保留一切可保留的古树；建筑与景观相结合，呈现葡萄串状空间形态；尊重并恢复原有的院落体系；尊重并恢复原有的街巷结构及空间尺度；尊重并恢复原有城市天际线；结合重要建筑，设计开放的城市空间；尽量保留完整的建筑外墙体；保留街区中具有宁波地方特色的建筑细部；保留原有的屋顶体系；重新设计并改造室内空间使其具有全新的丰富功能。

保留一切可保留的院落

项目采用原地保护和迁建保护的方式：对需要完全保留的重要历史建筑，采用原地保留、精心修缮、尽力保存历史记录的原则；对于部分拆除重建的建筑采用真实还原历史风貌，适当引入新功能的原则，使人们感受到历史和文化

的积淀。设计师根据前期的调查分析，将保留的院落分为两类：一类是完全保护院落，一类是保护加改造院落。

尊重并恢复原有的院落体系

院落是中国民居建筑的重要组成部分，南方院落和北方院落有区别，这与地理位置的差异有关。拆除后来搭建的建筑，把封闭在内的院落打开，恢复原有的空间，同时尽可能地发挥商业价值的要求，对外部空间进行功能组合和环境布置，使原有的建筑风貌和建筑装饰得以保留，创造更加宽敞的户外就餐和公众聚会的空间。

尊重并恢复原有街巷结构及空间尺度

历史街区的尺度大小是在历史中演变而来的，一般居住建筑街巷尺度较狭窄，形成了独特的空间形象。莲桥街历史街区的毛衙街、塔影巷等也是较狭窄。规划中，设计师在满足消防、集散等基础上，尽量保持原有街巷的尺度，还原历史。历史街区的建筑密度较高，设计中局部打开，使整个街巷空间有收有放，增加了街巷空间的趣味性，增强人们的舒适性。

尽量保留完整的建筑外墙体

设计尽可能保留了所有的建筑外墙体、建筑整

体外部空间和氛围，从而能够如实反应建筑风貌，并使保留建筑与周边的新建筑及空间相互衬托。立面和墙体形式是莲桥街街区重要历史特征，尽可能保留、加固，但不存在承重作用。

景观设计

全区以一条轴线、多个景观节点建立完整的空间体系，打造一条串联了人文、商业、景观的公共活动轴线，从北到南联系了天封塔、历史风貌区、延庆寺、旧城更新区和奉化江，构建了功能完善的城市片区。

恢复原有院落分析图

宁波莲桥街
Lianqiao Street, Ningbo

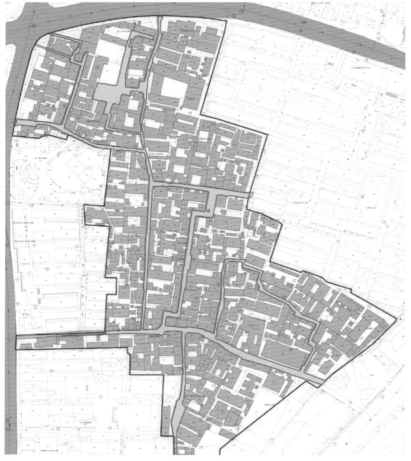

街巷现状分析图

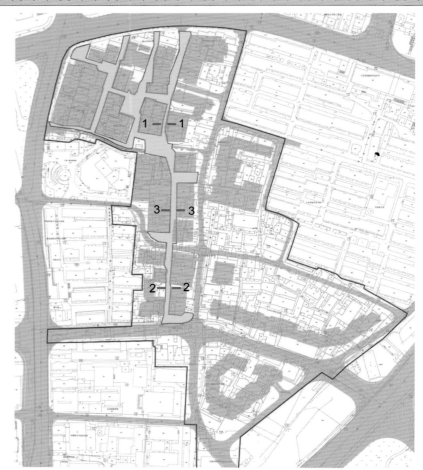

保留街巷分析图

The project gathers traditional alley styled street, boutique commerce, club, school, high-end apartment, etc. It is a typical urban case of historical street development and renovation. The project aims to build a high-end and comprehensive property for Ningbo center, and become an important milestone of Ningbo high-end residence development, and set up a new paradigm for ancient town renovation and historical site protection.

Planning Concept

Retain any retainable courtyards; respect existing texture and relocate some important courtyards; retain any retainable ancient trees; combine building with landscape to present a botryoid spatial form; respect and recover existing courtyard system; respect and recover existing alley structure and spatial scale; respect and recover existing city skyline; combine important buildings and design open urban space; keep outer wall intact as far as possible; retain architectural details of strong local features; retain existing roof system; redesign and renovate interior space to fill brand-new functions.

Retain Any Retainable Courtyards

There are two kinds of protection for buildings, one is retaining and the other is relocating. Important historical buildings that need to be retained intact are retained in their own seats, and are repaired carefully, so as to remain historical trace as far as possible. Some buildings that need to be dismantled and rebuilt are relocated and properly

added with new functions, and their historical style is recovered vividly, thus modern people can still feel the historical and cultural connotation. According to earlier research analysis, the retained courtyards are classified into only protection courtyards and protection plus renovation courtyards.

Respect and Recover Existing Courtyard System

Courtyard is an important component in Chinese residential buildings. There are differences between southern and northern courtyards due to their different geographical location. The dismantled and rebuilt buildings open their courtyards that were closed inside before; recover existing space; group functions and arrange environment in outside space, in view of maximizing commercial value; retain existing architectural style and decoration; create more spacious outdoor dining space and public gathering space.

Respect and Recover Existing Alley Structure and Spatial Scale

The spatial scale of the historical street features a gradual historical evolution. Normally, the street scale among residential buildings is quite small, forming unique space style. In the project, Maoya Street and Taying Valley are also very small in their spatial scale. During planning, except for meeting the requirements of fire protection, collection and evacuation, the design also tries to retain existing street scale as far as possible to recover history. Due to high building density, the design opens

some area, thus increase the interestingness of street space and improve comfortableness.

Keep Outer Wall Intact As Far As Possible

The design tries to retain all the outer walls, as well as the outer space and atmosphere of buildings, thus to recover the buildings as they were. Meanwhile, retained buildings contrast with surrounding buildings and space. The facades and walls are retained as much as possible, and are reinforced, but have no load-bearing function.

Landscape Design

The project builds a complete space system by an axis and multiple landscape nodes, creating a public activity axis that links up culture, commerce and landscape, and connects Tianfeng Tower, historical sites, Yanqing Temple, old town renovation area and Fenghua River from north to south.

占地面积：64 000平方米　建筑面积：118 184.8平方米　地点：中国浙江省宁波市
Site Area: 64,000m²　Building Area: 118,184.8m²　Location: Ningbo, Zhejiang Province, China

[解析商业街区与商业建筑设计]：商业街＆社区商业　　[Design Analysis of Commercial Districts & Commercial Buildings]：Commercial Streets & Community Commerce

宁波莲桥街
Lianqiao Street, Ningbo

占地面积：64 000平方米　建筑面积：118 184.8平方米　地点：中国浙江省宁波市
Site Area: 64,000m²　Building Area: 118,184.8m²　Location: Ningbo, Zhejiang Province, China

宁波莲桥街
Lianqiao Street, Ningbo

占地面积：64 000平方米　建筑面积：118 184.8平方米　地点：中国浙江省宁波市
Site Area: 64,000m²　Building Area: 118,184.8m²　Location: Ningbo, Zhejiang Province, China

武汉世界城光谷步行街
Smile at the World Pedestrian Street, Wuhan

开发商：武汉市世界城置业有限公司
Developer: Wuhan World Real Estate Co., Ltd.

设计公司：国内贸易工程设计研究院昌生建筑工作室　设计团队：陈昌生、任慧强、刘涛、张强、周伟强、由嘉、钱明光、凌明刚、冯静　占地面积：417 900平方米　建筑面积：1 800 000平方米　地点：中国湖北省武汉市
Design Company: Internal Trade Engineering Design & Research Institute Changsheng Architectural Design　Designers: Chen Changsheng, Ren Huiqiang, Liu Tao, Zhang Qiang, Zhou Weiqiang, You Jia, Qian Mingguang, Ling Minggang, Feng Jing
Site Area: 417,900m²　Building Area: 1,800,000m²　Location: Wuhan, Hubei Province, China

武汉世界城光谷步行街
Smile at the World Pedestrian Street, Wuhan

开发商：武汉市世界城置业有限公司
Developer: Wuhan World Real Estate Co., Ltd.

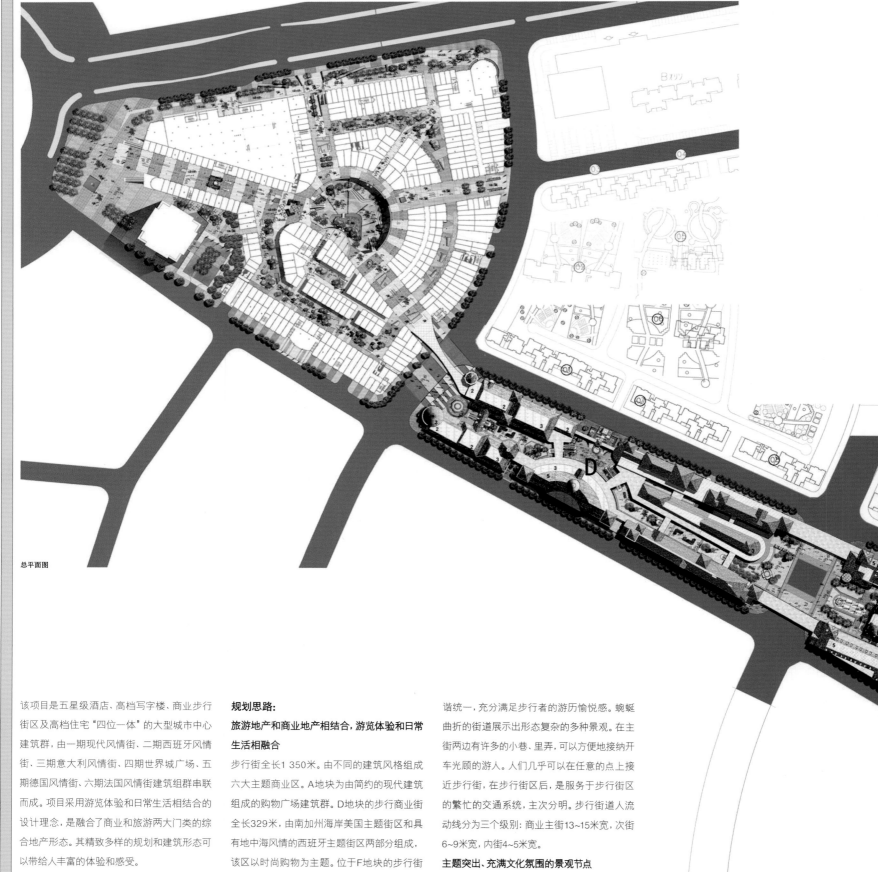

总平面图

该项目是五星级酒店、高档写字楼、商业步行街区及高档住宅"四位一体"的大型城市中心建筑群，由一期现代风情街、二期西班牙风情街、三期意大利风情街、四期世界城广场、五期德国风情街、六期法国风情街建筑组群串联而成。项目采用游览体验和日常生活相结合的设计理念，是融合了商业和旅游两大门类的综合地产形态。其精致多样的规划和建筑形态可以带给人丰富的体验和感受。

总建筑面积约1 800 000平方米，其中商业近800 000平方米，有10 000多个沿街商铺，可同时容纳3 000 000人在这里逛街，是武汉当前最大的商业地产项目之一。采用复合步行商业空间、大型专业卖场、购物中心、品牌专区的业态组合模式，涵盖了商业、娱乐、休闲体育、餐饮等，采用出售、租赁相合的经营形式。

规划思路：

旅游地产和商业地产相结合，游览体验和日常生活相融合

步行街全长1 350米。由不同的建筑风格组成六大主题商业区。A地块是由简约的现代建筑组成的购物广场建筑群。D地块的步行商业街全长329米，由南加州海岸美国主题街区和具有地中海风情的西班牙主题街区两部分组成，该区以时尚购物为主题。位于F地块的步行街全长427米，由意大利主题与德国主题步行街区两部分组成。意大利主题步行街区以婚庆为商业主题。德国主题步行街区以酒吧文化及餐饮文化为主题。位于H地块的步行商业街全长200米，为法国主题街区，其定位为住宅、办公、酒店等综合性商业区。

创造充满探索性的商业动线与多极化商业空间

在街道和广场的设计中把握对称与非对称的和谐统一，充分满足步行者的游历愉悦感。蜿蜒曲折的街道展示出形态复杂的多种景观。在主街两边有许多的小巷、里弄，可以方便地接纳开车光顾的游人。人们几乎可以在任意的点上接近步行街，在步行街区后，是服务于步行街区的繁忙的交通系统，主次分明。步行街道人流动线分为三个级别：商业主街13~15米宽，次街6~9米宽，内街4~5米宽。

主题突出、充满文化氛围的景观节点

本方案共设计十大主题景观节点。每600~800米设置标志性广场，起到重新组织人流的作用，作为逛街的起点和终点，同时具有节庆、聚会、表演的功能。每80~100米设置节点广场，起到空间变化和纵向人流组织的作用。600~800米是人在一次性逛街的过程中的极限距离。其标志性意义相当重要，本项目分别在东西两端和中部设置八个标志性景观节点。

设计公司：**国内贸易工程设计研究院昌生建筑工作室** 设计团队：**陈昌生、任慧强、刘涛、张强、周伟强、由嘉、钱明光、凌明刚、冯静** 占地面积：**417 900平方米** 建筑面积：**1 800 000平方米** 地点：**中国湖北省武汉市**
Design Company: Internal Trade Engineering Design & Research Institute Changsheng Architectural Design　Designers: Chen Changsheng, Ren Huiqiang, Liu Tao, Zhang Qiang, Zhou Weiqiang, You Jia, Qian Mingguang, Ling Minggang, Feng Jing
Site Area: 417,900m²　Building Area: 1,800,000m²　Location: Wuhan, Hubei Province, China

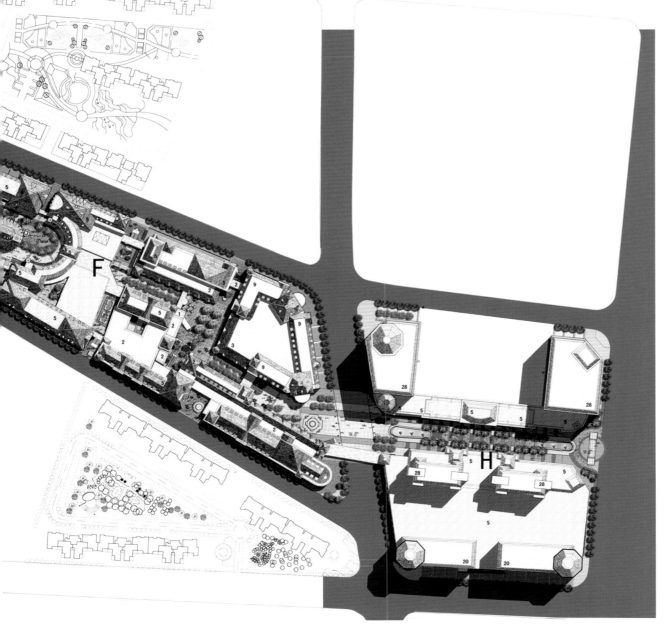

建筑设计

建筑采用新古典主义风格，设计上把握不同分区的建筑特点，运用现代材料和工艺来传承欧式古典建筑的美感，着力打造多样梦幻的游览体验。一期现代风情街外立面采用深色玻璃幕墙结合干挂石材、铝板等材质，建筑形体对比强烈、大开大合，整体风格精致冷峻、独树一帜；二期西班牙风情街建筑外立面采用文化石和仿石漆相结合，平缓的四坡屋顶配合深红色陶瓦，结合多层次的退台空间，整体氛围轻松、休闲；三期意大利风情街建筑外立面采用干挂石材、墙身的转角采用石柱装饰，檐下的线脚、建筑转角的八角形塔楼、教堂的尖顶、外廊的拱圈，处处精雕细刻，无一不体现出意大利建筑的热情与浪漫；四期世界城广场和五期德国风情街采用德式城堡风格，外墙采用深驼色干挂石材，厚重的建筑形体、高耸的屋顶和塔楼将德国城堡威严雄伟的格调表现得淋漓尽致；六期法国风情街采用对称建筑造型，外立面色彩典雅、清新，细节处理上，在屋顶运用精致的老虎窗，经典的法式廊柱、雕花、线条等元素，处处体现出法式古典建筑的精致优雅。

[解析商业街区与商业建筑设计]：商业街＆社区商业　　[Design Analysis of Commercial Districts & Commercial Buildings]：Commercial Streets & Community Commerce

武汉世界城光谷步行街
Smile at the World Pedestrian Street, Wuhan

开发商：武汉市世界城置业有限公司
Developer: Wuhan World Real Estate Co., Ltd.

设计公司: 国内贸易工程设计研究院昌生建筑工作室　设计团队: 陈昌生、任慧强、刘涛、张强、周伟强、由嘉、钱明光、凌明刚、冯静　占地面积: 417 900平方米　建筑面积: 1 800 000平方米　地点: 中国湖北省武汉市
Design Company: Internal Trade Engineering Design & Research Institute Changsheng Architectural Design　Designers: Chen Changsheng, Ren Huiqiang, Liu Tao, Zhang Qiang, Zhou Weiqiang, You Jia, Qian Mingguang, Ling Minggang, Feng Jing
Site Area: 417,900m²　Building Area: 1,800,000m²　Location: Wuhan, Hubei Province, China

The project is a huge building cluster in the city center, housing five-star hotel, high-end office building, commercial walking street and high-end residence. It is composed of six phases including phase 1 modern style street, phase 2 Spanish style street, phase 3 Italian style street, phase 4 world plaza, phase 5 Germany style street and phase 6 French style street. It features a design concept that combines tour experience and daily life, becoming a complex property that mixes commerce and tourism. It is worth mentioning that the diverse planning and building shape will definitely bring people with rich new experiences.

The project has a gross floor area of 1,800,000m², among which the commercial area captures 800,000m². As one of Wuhan's largest commercial properties, the project has more than 10,000 street stores, which can accommodate 3,000,000 people. It is equipped with composite walking commercial space, large exclusive mall, shopping center and brand zone, involving commerce, entertainment, sports, and F&B. Besides, the project adopts an operation mode that combines partial sale and partial lease.

Planning Concept

Combining tourism property with commercial property, and tour experience with daily life:

The walking street is 1,350m long. The six theme commercial zones differ from each other in their building style. Land A is a shopping plaza spread with simple modern-style buildings. Land D has a 329m long walking street which is made of American style street and Spanish style street, and features a theme of fashion shopping. The 427m long walking street on Land F is formed by Italian

style street and Germany style street, of which, the former features wedding as them, while the latter boasts bar culture and F&B culture as theme. The 200m long French-style walking street on Land H aims to be a service commercial area covering residence, office and hotel.

Creating Exploratory Commercial Circulation and Multipolar Commercial Space:

In the design of street and square, symmetry and asymmetry are in harmony and unity, increasing people's joyousness when walking through. Various landscapes come into sight as people walking along the winding street. There are many alleys on the sides of the main street, facilitating tourists who drive. People can access the walking street at any spots. Behind the walking street is a busy traffic system with clear priorities. The pedestrian circulation for the walking street is classified into three levels: 13-15m wide main street, 6-9m wide secondary street and 4-5m wide inner street.

Theme-prominent and Culture-filled Landscape

The project boasts ten theme landscape nodes. Landmark square is set at the interval of 600-800m to regroup customer flow, and act as the start point and end point of shopping, and serve festival celebration, party and show. Node square is placed at the interval of 80-100m to transform space and organize longitudinal customer flow. In a one-time shopping, 600-800m is people's maximum distance. The project sets eight landmark landscape nodes at the east and west ends and middle part.

Architecture Design

The buildings feature neoclassical style. According to the specific architectural characteristics of

different lands, the design uses modern materials and skills in the purpose of inheriting the aesthetics of European classic architecture and creating diverse and dreamy tour experience. The facade of phase 1 modern style street combines dark-tone glass curtain wall with dry-hang stone and aluminum sheet, forming sharp contrast on the building shape. The facade of phase 2 Spanish style street combines culture stone with stone paint. Flat hipped roof is combined with dark red terra-cotta tile. Multi-layer set-back spaces are introduced. The result is a relaxed and leisurely style. Phase 3 Italian style street uses dry-hang stone on building facade, and stone pillar on wall corners. The architrave below eave, octagonal tower at building corner, pinnacle of church, arch ring of outer corridor, carving everywhere, all of these tell the passion and romance of Italian architecture. Phase 4 world plaza and phase 5 Germany style street are of Germany castle style, with outer wall adopting dark camel dry-hang stone, as well as heavy building shape, towering roof and tower, all of which incisively embody the grandness and majesty of Germany castle. Phase 6 French style street introduces symmetrical building shape, stained with elegant and fresh colors. In the detail design, exquisite dormers on the roof, classical French portico, carving, lines, etc. can be commonly seen, vividly revealing the charm and beauty of French classical architecture.

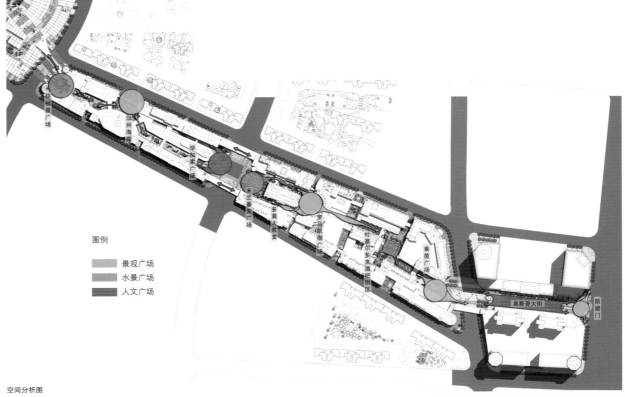

图例

景观广场
水景广场
人文广场

空间分析图

武汉世界城光谷步行街
Smile at the World Pedestrian Street, Wuhan

开发商：武汉市世界城置业有限公司
Developer: Wuhan World Real Estate Co., Ltd.

区位分析图

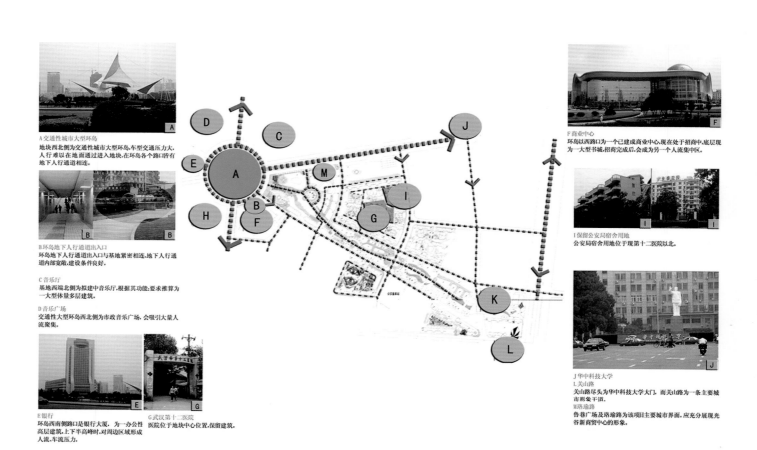

A 交通性城市大型环岛
地块西北侧为交通性城市大型环岛，车型交通压力大，人行难以在地面通过进入地块，在环岛各个路口皆有地下人行通道相连。

B 环岛地下人行通道出入口
环岛地下人行通道出入口与基地紧密相连，地下人行通道内部宽敞，建设条件良好。

C 音乐厅
基地西端北侧为拟建中音乐厅，根据其功能，要求推算为一大型体量多层建筑。

D 音乐广场
交通性大型环岛西北侧为市政音乐广场，会吸引大量人流聚集。

E 银行
环岛西南侧路口是银行大厦，为一办公性高层建筑，上下半高峰时，对周边区域形成人流、车流压力。

G 武汉第十二医院
医院位于地块中心位置，保留建筑。

F 商业中心
环岛以西路口为一个已建成商业中心，现在处于招商中，底层现为一大型书城，招商完成后，会成为另一个人流集中区。

I 保留公安局宿舍用地
公安局宿舍用地位于现第十二医院以北。

J 华中科技大学
L 关山路
关山路尽头为华中科技大学大门，而关山路为一条主要城市形象干道。

M 路瑜路
鲁巷广场及珞瑜路为该项目主要城市界面。应充分展现光谷新商贸中心的形象。

现状分析图

设计公司：国内贸易工程设计研究院昌生建筑工作室　设计团队：陈昌生、任慧强、刘涛、张强、周伟强、由嘉、钱明光、凌明刚、冯静　占地面积：417 900平方米　建筑面积：1 800 000平方米　地点：中国湖北省武汉市
Design Company: Internal Trade Engineering Design & Research Institute Changsheng Architectural Design　Designers: Chen Changsheng, Ren Huiqiang, Liu Tao, Zhang Qiang, Zhou Weiqiang, You Jia, Qian Mingguang, Ling Minggang, Feng Jing
Site Area: 417,900m²　Building Area: 1,800,000m²　Location: Wuhan, Hubei Province, China

武汉世界城光谷步行街
Smile at the World Pedestrian Street, Wuhan

开发商: 武汉市世界城置业有限公司
Developer: Wuhan World Real Estate Co., Ltd.

设计公司: 国内贸易工程设计研究院昌生建筑工作室　设计团队: 陈昌生、任慧强、刘涛、张强、周伟强、由嘉、钱明光、凌明刚、冯静　占地面积: 417 900平方米　建筑面积: 1 800 000平方米　地点: 中国湖北省武汉市
Design Company: Internal Trade Engineering Design & Research Institute Changsheng Architectural Design　Designers: Chen Changsheng, Ren Huiqiang, Liu Tao, Zhang Qiang, Zhou Weiqiang, You Jia, Qian Mingguang, Ling Minggang, Feng Jing
Site Area: 417,900m²　Building Area: 1,800,000m²　Location: Wuhan, Hubei Province, China

武汉世界城光谷步行街
Smile at the World Pedestrian Street, Wuhan

开发商：武汉市世界城置业有限公司
Developer: Wuhan World Real Estate Co., Ltd.

设计公司: 国内贸易工程设计研究院昌生建筑工作室　设计团队: 陈昌生、任慧强、刘涛、张强、周伟强、由嘉、钱明光、凌明刚、冯静　占地面积: 417 900平方米　建筑面积: 1 800 000平方米　地点: 中国湖北省武汉市
Design Company: Internal Trade Engineering Design & Research Institute Changsheng Architectural Design　Designers: Chen Changsheng, Ren Huiqiang, Liu Tao, Zhang Qiang, Zhou Weiqiang, You Jia, Qian Mingguang, Ling Minggang, Feng Jing
Site Area: 417,900m²　Building Area: 1,800,000m²　Location: Wuhan, Hubei Province, China

武汉世界城光谷步行街
Smile at the World Pedestrian Street, Wuhan

开发商：武汉市世界城置业有限公司
Developer: Wuhan World Real Estate Co., Ltd.

设计公司: 国内贸易工程设计研究院昌生建筑工作室　设计团队: 陈昌生、任慧强、刘涛、张强、周伟强、由嘉、钱明光、凌明刚、冯静　占地面积: 417 900平方米　建筑面积: 1 800 000平方米　地点: 中国湖北省武汉市
Design Company: Internal Trade Engineering Design & Research Institute Changsheng Architectural Design　Designers: Chen Changsheng, Ren Huiqiang, Liu Tao, Zhang Qiang, Zhou Weiqiang, You Jia, Qian Mingguang, Ling Minggang, Feng Jing
Site Area: 417,900m²　Building Area: 1,800,000m²　Location: Wuhan, Hubei Province, China

武汉世界城光谷步行街
Smile at the World Pedestrian Street, Wuhan

开发商：武汉市世界城置业有限公司
Developer: Wuhan World Real Estate Co., Ltd.

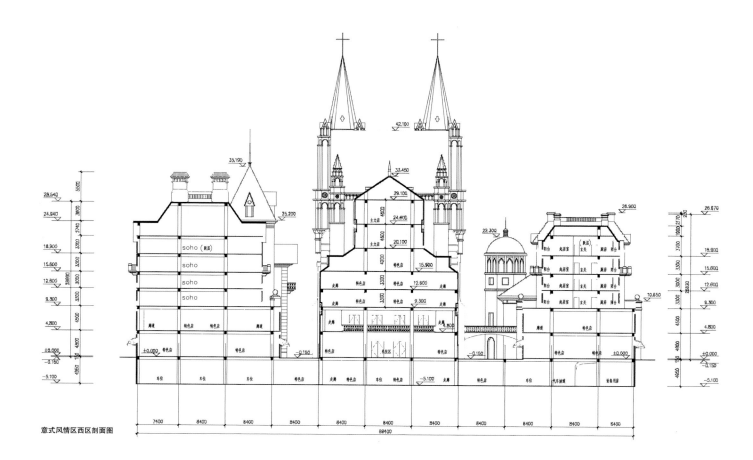

意式风情区西区剖面图

设计公司：国内贸易工程设计研究院昌生建筑工作室　设计团队：陈昌生、任慧强、刘涛、张强、周伟强、由嘉、钱明光、凌明刚、冯静　占地面积：417 900平方米　建筑面积：1 800 000平方米　地点：中国湖北省武汉市
Design Company: Internal Trade Engineering Design & Research Institute Changsheng Architectural Design　Designers: Chen Changsheng, Ren Huiqiang, Liu Tao, Zhang Qiang, Zhou Weiqiang, You Jia, Qian Mingguang, Ling Minggang, Feng Jing
Site Area: 417,900m²　Building Area: 1,800,000m²　Location: Wuhan, Hubei Province, China

武汉世界城光谷步行街
Smile at the World Pedestrian Street, Wuhan

开发商: 武汉市世界城置业有限公司
Developer: Wuhan World Real Estate Co., Ltd.

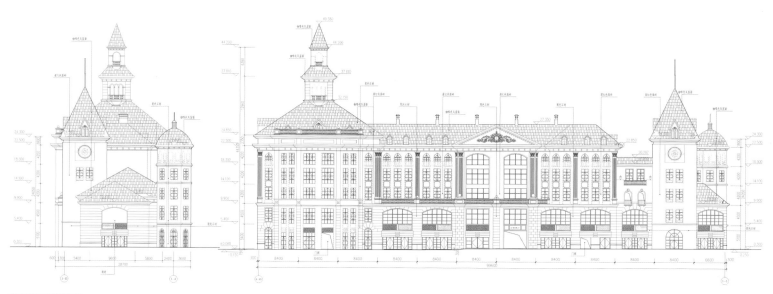

意式风情区西区立面图

设计公司: 国内贸易工程设计研究院昌生建筑工作室　设计团队: 陈昌生、任慧强、刘涛、张强、周伟强、由嘉、钱明光、凌明刚、冯静　占地面积: 417 900平方米　建筑面积: 1 800 000平方米　地点: 中国湖北省武汉市
Design Company: Internal Trade Engineering Design & Research Institute Changsheng Architectural Design　Designers: Chen Changsheng, Ren Huiqiang, Liu Tao, Zhang Qiang, Zhou Weiqiang, You Jia, Qian Mingguang, Ling Minggang, Feng Jing
Site Area: 417,900m²　Building Area: 1,800,000m²　Location: Wuhan, Hubei Province, China

武汉世界城光谷步行街
Smile at the World Pedestrian Street, Wuhan

开发商: 武汉市世界城置业有限公司
Developer: Wuhan World Real Estate Co., Ltd.

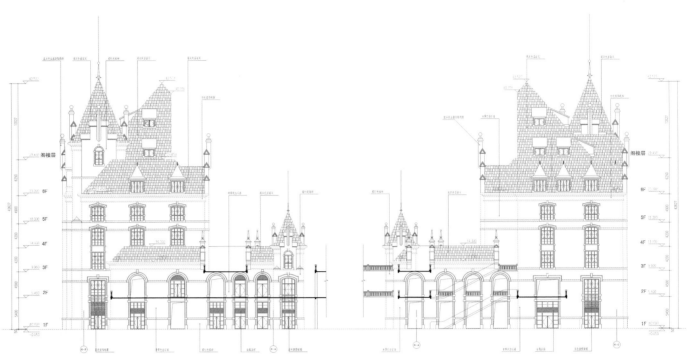

德式立面图

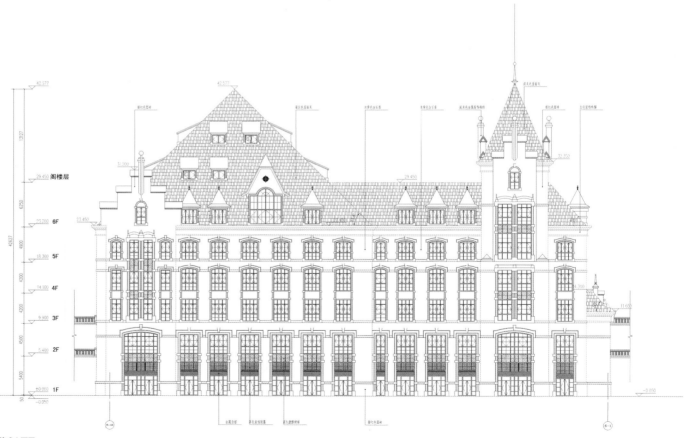

德式立面图

设计公司：国内贸易工程设计研究院昌生建筑工作室　设计团队：陈昌生、任慧强、刘涛、张强、周伟强、由嘉、钱明光、凌明刚、冯静　占地面积：417 900平方米　建筑面积：1 800 000平方米　地点：中国湖北省武汉市
Design Company: Internal Trade Engineering Design & Research Institute Changsheng Architectural Design　Designers: Chen Changsheng, Ren Huiqiang, Liu Tao, Zhang Qiang, Zhou Weiqiang, You Jia, Qian Mingguang, Ling Minggang, Feng Jing
Site Area: 417,900m²　Building Area: 1,800,000m²　Location: Wuhan, Hubei Province, China

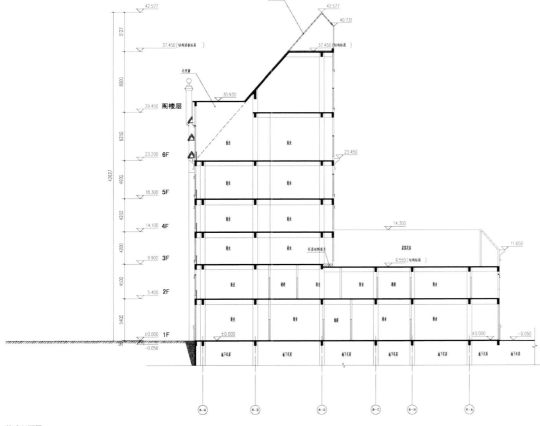

德式剖面图

武汉世界城光谷步行街
Smile at the World Pedestrian Street, Wuhan

开发商：武汉市世界城置业有限公司
Developer: Wuhan World Real Estate Co., Ltd.

设计公司: 国内贸易工程设计研究院昌生建筑工作室   设计团队: 陈昌生、任慧强、刘涛、张强、周伟强、由嘉、钱明光、凌明刚、冯静   占地面积: 417 900平方米   建筑面积: 1 800 000平方米   地点: 中国湖北省武汉市
Design Company: Internal Trade Engineering Design & Research Institute Changsheng Architectural Design   Designers: Chen Changsheng, Ren Huiqiang, Liu Tao, Zhang Qiang, Zhou Weiqiang, You Jia, Qian Mingguang, Ling Minggang, Feng Jing
Site Area: 417,900m²   Building Area: 1,800,000m²   Location: Wuhan, Hubei Province, China

[解析商业街区与商业建筑设计]：商业街＆社区商业　　Design Analysis of Commercial Districts & Commercial Buildings：Commercial Streets & Community Commerce

合肥1912
1912, Hefei

开发商：金大地集团　建筑设计：上海日清建筑设计有限公司　占地面积：15 367.8平方米　建筑面积：83 452平方米　容积率：1.2　绿化率：40.1%　地点：中国安徽省合肥市黄山路怀宁路交口

Developer: Golden Land Group　Architecture Design: Shanghai Lacime Architectural Design Co., Ltd.　Site Area: 15,367.8m²　Building Area: 85,452m²　Plot Ratio: 1.2　Greening Rate: 40.1%　Location: the intersection of Huangshan Road and Huaining Road, Hefei, Anhui Province, China

合肥1912
1912, Hefei

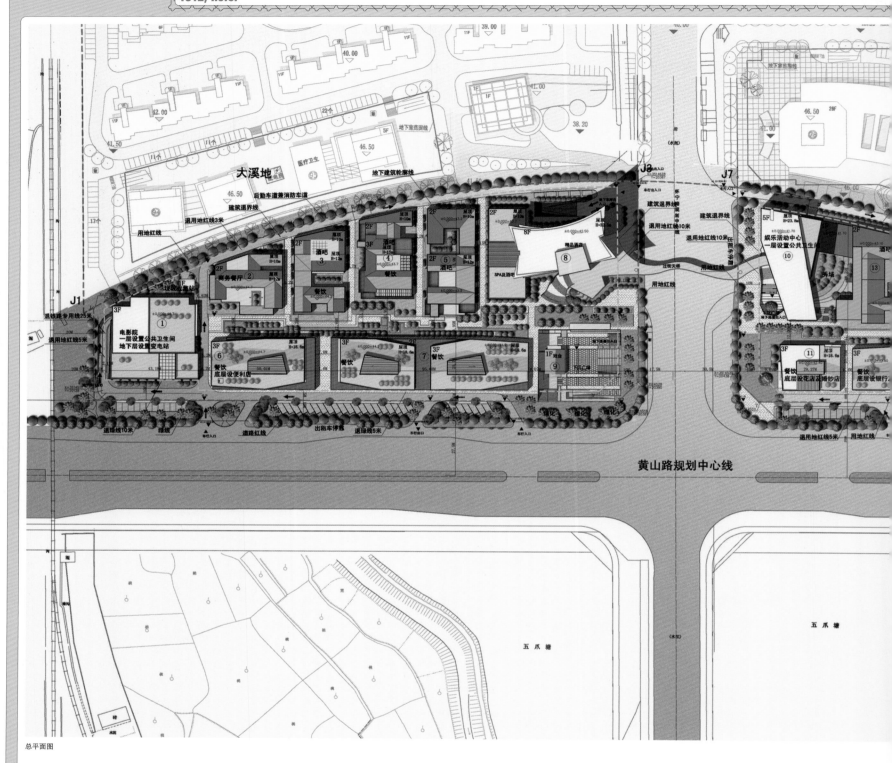

总平面图

本项目定位为合肥市"城市的客厅""城市的名片"，肩负着提升合肥城市消费层次，激励城市职能向第三产业更快、更好地转移的重任，同时，对改善黄山路沿线及整个政务区、高新区的城市面貌，提高商业及居住品质有着直接促进作用。项目从着手规划至全面开街营业，仅用了21个月的时间。

在空间处理上，设计师以一条主题商业内街为主轴，将沿黄山路的沿街高档餐饮区和内部的情景合院式的酒吧餐饮区串联，1号楼电影院、8号楼精品酒店、10号楼娱乐中心及16号楼经济型酒店穿插其中，以其较为醒目的形体形成地标，亦成为情景商业的绚丽背景。内街和建筑，建筑和城市界面之间，形成多个主题庭院和广场，结合鱼骨状的传统街巷空间，营造出极具气氛的室外商业场所。在最核心的沿怀宁路两侧的内街入口处，西侧地块的下沉庭院和古戏楼及东侧地块的黄梅戏院，形成视觉中心、时空的节点。历史与明天，文化与商业在此邂逅，和谐而统一。

作为商业建筑，单体设计遵循普适性原则与量身定做相结合的方式，很好地满足了业主对使用功能既求灵活可变，又需满足业态配比，保证重点商铺的要求。同时，设计注重对交流空间的营造和关怀，各具情趣的退台、露台、庭院，促进了人与人的交往，也提升了该区域的商业价值。

设计师力求在项目上创新，带给人们一个有着文化气氛的精细的观感体验。非标节点的U玻、波形钢板、金属瓦屋面、隐蔽式通风百叶、点支玻璃幕墙等，无不经过反复推敲和深化，并最终以幕墙使用说明书的形式作为招投标文件，对幕墙施工起到指导作用。

开发商: 金大地集团   建筑设计: 上海日清建筑设计有限公司   占地面积: 15 367.8平方米   建筑面积: 83 452平方米   容积率: 1.2   绿化率: 40.1%   地点: 中国安徽省合肥市黄山路怀宁路交口
Developer: Golden Land Group   Architecture Design: Shanghai Lacime Architectural Design Co., Ltd.   Site Area: 15,367.8m²   Building Area: 85,452m²   Plot Ratio: 1.2   Greening Rate: 40.1%   Location: the intersection of Huangshan Road and Huaining Road, Hefei, Anhui Province, China

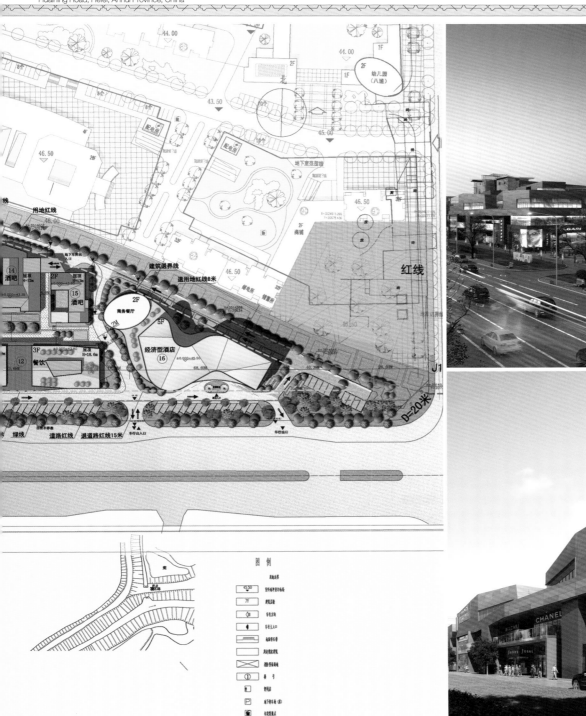

As the city parlor and city card of Hefei, the project shoulders an important mission: to upgrade the consumption standard of Hefei and to encourage faster and better transformation of the urban function into tertiary industry. Needless to say, the project directly beautifies the look of Huangshan Road, the entire governmental affair district and high-tech district, energizes commerce and improves residential quality.

As for the layout, the project features an theme commercial inner street as the main axis, connecting the high-class F&B zone along Huangshan Road with the inner courtyard-style F&B zone, and meanwhile with cinema building,

boutique hotel building, entertainment building and economic hotel building interspersed in, all of which, with eye-catching shape, not only form a landmark, but also become an amazing background of the scenic commerce. Multiple theme courtyards and squares are formed between inner street and building, building and city interface, which, when combined with herringbone traditional alley space, creates a featured exterior commercial space. At the most essential entrances of the inner street which are set at the two sides of Huaining Road, the sunken courtyard and ancient theater in the west site, together with the Huangmei Opera theater in the east site, form the visual focus.

History and future, culture and commerce meet here and are harmoniously united as one.

As a commercial building, the project design pays attention to both universality and uniqueness, meeting the client's requirement of flexible and changeable functions that satisfy business arrangement and ensure key stores. The design also lays emphasis on people's communication space. For example, all kinds of terraces and courtyards are created to promote social communication among people and to level up the commercial value.

The project continues many mature node practices and materials that are quite commonly seen in

Lacime. It also strives to be innovative, so as to obtain an experience which is fully filled with cultural atmosphere. Non-node U-shape glass, corrugated steel sheet, metal tile roofing, concealed ventilation shutters and point-supported glass rib curtain wall, all of these are analyzed again and again.

合肥1912
1912, Hefei

开发商：金大地集团　建筑设计：上海日清建筑设计有限公司　占地面积：15 367.8平方米　建筑面积：83 452平方米　容积率：1.2　绿化率：40.1%　地点：中国安徽省合肥市黄山路怀宁路交口
Developer: Golden Land Group　Architecture Design: Shanghai Lacime Architectural Design Co., Ltd.　Site Area: 15,367.8m²　Building Area: 85,452m²　Plot Ratio: 1.2　Greening Rate: 40.1%　Location: the intersection of Huangshan Road and Huaining Road, Hefei, Anhui Province, China

合肥1912
1912, Hefei

开发商：金大地集团　建筑设计：上海日清建筑设计有限公司　占地面积：15 367.8平方米　建筑面积：83 452平方米　容积率：1.2　绿化率：40.1%　地点：中国安徽省合肥市黄山路怀宁路交口
Developer: Golden Land Group　Architecture Design: Shanghai Lacime Architectural Design Co., Ltd.　Site Area: 15,367.8m²　Building Area: 85,452m²　Plot Ratio: 1.2　Greening Rate: 40.1%　Location: the intersection of Huangshan Road and Huaining Road, Hefei, Anhui Province, China

合肥1912
1912, Hefei

开发商：金大地集团　建筑设计：上海日清建筑设计有限公司　占地面积：15 367.8平方米　建筑面积：83 452平方米　容积率：1.2　绿化率：40.1%　地点：中国安徽省合肥市黄山路怀宁路交口
Developer: Golden Land Group　Architecture Design: Shanghai Lacime Architectural Design Co., Ltd.　Site Area: 15,367.8m²　Building Area: 85,452m²　Plot Ratio: 1.2　Greening Rate: 40.1%　Location: the intersection of Huangshan Road and Huaining Road, Hefei, Anhui Province, China

合肥1912
1912, Hefei

开发商: 金大地集团　建筑设计: 上海日清建筑设计有限公司　占地面积: 15 367.8平方米　建筑面积: 83 452平方米　容积率: 1.2　绿化率: 40.1%　地点: 中国安徽省合肥市黄山路怀宁路交口
Developer: Golden Land Group　Architecture Design: Shanghai Lacime Architectural Design Co., Ltd.　Site Area: 15,367.8m²　Building Area: 85,452m²　Plot Ratio: 1.2　Greening Rate: 40.1%　Location: the intersection of Huangshan Road and Huaining Road, Hefei, Anhui Province, China

合肥1912
1912, Hefei